URBAN QUAIL-KEEPING

Karen J Puddephatt

URBAN QUAIL-KEEPING

Karen J Puddephatt

www.UrbanQuailKeeping.com

Urban Quail-Keeping
First published in 2014 by
Lemon Tree Farm Press
17 Acremead, Warmington, Peterborough, PE8 6TP, United Kingdom
Tel: (01832) 280949
Email: lemontreefarmpress@yahoo.co.uk
Website: www.UrbanQuailKeeping.com

Acknowledgments

I would like to thank all the quail keepers along the way who have given me numerous quail-keeping hints and tips and pointed me in the right direction both in person and online. I would like to thank Alex Puddephatt, for his wonderful photographs, editing help, endless patience and cups of tea, over the duration of writing this book. Finally I would like to thank the Humane Slaughter Association, for their helpful insights and advice.

Disclaimer

The contents of this book are for the intention of those people intending to keep quail for a hobby. The author accepts no responsibility for any loss or consequential loss as a result of following any advice contained in the book.

All rights reserved. No part of this book may be reproduced or transmitted in any form or by any means, electronic or mechanical, including photocopying, recording, or by any information storage and retrieval system, without written permission from the copyright owner.

All photographs are by the author and Alex Puddephatt.

Copyright © 2014 Karen J Puddephatt

All rights reserved.

ISBN-13: 978-1484030028

CONTENTS

CHAPTER 1
EGG PRODUCTION — 9
- How Many Eggs Do Quail Lay? — 10
- How Many Quail Do I Need? — 11
- Which Strains of Quail to Buy? — 11
- Common Quail Breeds for Egg and Meat Production — 12
- What Do Quail's Eggs Taste Like? — 13
- Eggshell Colouring — 14
- Odd Sized Eggs — 14
- A Tip for Egg Collection — 17
- Selling Quail Eggs — 17

CHAPTER 2
QUINTESSENTIALLY QUAIL — 19
- How to Hold Your Quail — 20
- Taming Quail — 22
- Will Quail Make a Noise or Disturb My Neighbours? — 23
- Classification of Quail — 23
- The Thirteen Quail Genera — 24

CHAPTER 3
QUAIL HOUSING — 25
- I've Not Much Space So Where Can I Keep Quail? — 25
- Hutches — 26
- Runs — 26
- Can I Free Range Quail? — 26
- Basic Housing Requirements — 27
- Six Elements of Great Quail Housing — 27
- Hutch Designs — 30
- Seven Essential Items for Inside the Quail Hutch — 34

CHAPTER 4
QUAIL FOOD — 39
- Recommended Nutritional Requirements of Quail — 40
- Chart Showing Weight Gain of Two Quail — 41
- The Role of Nutrients — 41
- How Much Food Do Quail Need & What is the Cost? — 42
- Favourite Snacks and Treats — 42
- Foods to Avoid — 43

CHAPTER 5
QUAIL CARE — 44
- Daily Jobs — 44
- Weekly — 45
- Monthly — 45
- Six Monthly — 45
- Cleaning the Housing — 45

CHAPTER 6
PECKING PROBLEMS RESOLVED — 47
- Factors Causing Pecking — 48
- Reducing Pecking in Quail — 48

CHAPTER 7
HEALTH — 51
- A Hospital Cage — 52
- Beak and Toenail Health — 52
- Prolapsed Vent — 58
- How to Treat a Prolapse — 58
- Toeballing — 60
- Bumblefoot — 61
- Red Mites (*Dermanyssus gallinae*) — 61
- Scaly Leg Mite (*Knemidocoptes mutans*) — 62
- Lice — 62
- A Comparison to Distinguish Between Lice and Mites — 63
- Worms — 63
- Coccidiosis — 64
- Ulcerative Enteritis (Quail Disease) — 65
- Cuts and Abrasions — 65

CHAPTER 8
HOME CRAFT — 67
- Amazing Apple Cider Vinegar — 67
- The Uses of ACV — 68
- Precautions — 70
- The Correct Dosage of ACV — 70
- How to Make ACV — 70
- Breeding Mealworms — 73
- How to Deepen Egg Yolk Colour — 75
- Using Eggshells — 75

CHAPTER 9
MAKING MORE QUAIL — 77
   Breeding Basics — 77
   Obtaining Fertile Eggs — 80
   Handling Fertilized Eggs — 80
   Washing Eggs — 80
   Incubators — 81
   The Process of Incubation — 82
   Humidity — 82
   Candling — 84
   Pipping — 84
   Hatching — 85
   A Chick's First Hours — 85
   Moving your Chicks to the Brooder — 86
   Quail Incubation Check List — 87
   Brooders — 88
   Brooder Temperature — 90
   Ventilation — 91
   Growth — 93
   Bedding Material in the Early Days — 93
   Cleanliness — 94
   Nutritional Requirements of Quail Chicks — 94
   Appropriate Food Size — 94
   Nursery Pens — 95
   Moving Chicks to Their Outside Homes — 95
   Making a Plan for Your Male Chicks — 96
   Sexing Quail — 96
   Feather Colour and Genetic Dominance — 101
   Colouring: Dominant and Recessive Genes of Quail — 101
   Feather Mutations — 102

CHAPTER 10
KEEPING QUAIL FOR MEAT — 103
   The Benefits of Producing Your Own Meat — 103
   Selecting Birds to Eat — 104
   'When They Start to Crow, It's Time to Go!' — 104
   Breeds for the Table — 104
   How to Achieve a Constant Supply of Meat — 105
   The Eggs to Meat Calculation — 105
   Selection for Breeding — 105
   Light Restriction — 106
   What Age to Cull? — 106
   Selling Quail Meat — 106

CHAPTER 11
CULLING — 107
- Principles for a Painless Cull — 108
- The Best Culling Methods Without a Stunner — 108
- The Moments After Culling — 109
- The Location of the Cull — 109
- Up-to-Date Advice — 110
- Tips — 110

CHAPTER 12
PREPARING QUAIL FOR THE TABLE — 111
- How to Process the Carcass — 111
- Plucking — 111
- Skinning the Quail — 112
- Removing the Innards — 112

CHAPTER 13
COOKING WITH QUAIL EGGS — 113
- How to Crack Quail Eggs — 113
- How Long Does It Take to Hard Boil a Quail Egg? — 114
- How Long to Boil an Egg to Achieve a Runny Yolk? — 114
- Easy Methods for Shelling Quail Eggs — 114

CHAPTER 14
QUAIL EGG RECIPES — 116

CHAPTER 15
QUAIL MEAT RECIPES — 128

Appendices
- Appendix A: OLD WORLD QUAIL — 135
- Appendix B: NEW WORLD QUAIL — 137
- Appendix C: BUTTONQUAIL — 142
- Further Information — 145
- Index — 146
- About The Author — 151

CHAPTER 1

EGG PRODUCTION

If you live with limited space and would love fresh eggs every morning, but would like to keep poultry that is both easier than chickens and needing far less room, then *Urban Quail-Keeping* is your obvious choice. You can keep these enchanting birds in a small space such as a patio or balcony, garden or garage and they will provide you with countless eggs that are both nutritious and delicious. With a little help from my book you may quite easily produce your own delicious quail meat, a known delicacy. However it does not stop there as these delightful birds have a delicate subtle beauty and curious character.

With relatively little effort, following the advice given in these pages, you can set up your very own quail covey, providing optimal conditions for the health and happiness of your quail while benefiting from greater control over the quality and ethical nature of its production.

How Many Eggs Do Quail Lay?

Quail are prolific egg layers, with each bird laying at least one egg a day. Imagine fresh eggs every day, produced easily from your own home and at a very modest price. Egg production starts early as quail start laying at around only 35-45 *days* old as compared to 18 to 22 *weeks* for chickens and are into full production by the time they are only 50 days old.

Over the course of their prolific gourmet egg laying lives, quail can easily produce 300 eggs per year under good conditions and when provided with artificial lighting in the winter months. You can soon begin to produce lots of nutritious eggs in return for minimal effort.

Most of their eggs are laid in the afternoon and evenings, when you can hear them emit a cheering, triumphant crow. They seem very pleased with the result! With the additional benefit of being easy to maintain in either an urban or rural location they are the ideal bird for the home farmer.

A selection of Jumbo quail eggs

How Many Quail Do I Need?

I was once asked how many quail are needed to get the equivalent of a dozen 'normal' chicken's eggs per week. What a practical question! Opinions will vary as many factors come into play such as nutrition, housing, lighting and stress factors, all of which influence egg production, so I can only advise a very rough estimate.

A 'baker's dozen' of quail (i.e. 13) will lay the equivalent of a dozen chicken's eggs per week on average – over the course of their lives.

The latter part of this statement is as important as the first as your quail will lay more eggs up to around nine months to a year old and then reduce output.

Are you ready for some quail mathematics? To demonstrate how I got to this number the first thing to know is:

approximately five quail eggs = one medium chicken's egg.

Thirteen quail laying one egg each per day over seven days in the week will give you 91 quail eggs per week. When you divide this by five you will get the comparative number of chicken's eggs which is 18. Obviously this is the equivalent of one-and-a-half dozen chicken's eggs and not one dozen as indicated because as mentioned at first the birds will lay more eggs and then later significantly reduce their output. Also there may be losses and fatalities and escapees. The average egg production over the course of their lives is around 75% and 75% of our 91 quail eggs is approximately 68. This gives us our magic number equating to one dozen chicken's eggs. If you are all superstitious about the number 13 then you can always acquire more quail! If you are unable to use all of them yourself I'm sure relatives and friends will come to your assistance and I always like to have too many rather than too few.

Which Strains of Quail to Buy?

Your choice of bird will depend upon whether you want to keep them for egg or meat production or maybe both. Japanese quail are the basis for most of the egg and meat producing birds. As there is no official registration for breeding or naming, you may encounter variations in the names and colour combinations attributed to them and sizes may vary.

Common Quail Breeds for Egg and Meat Production

Breed	Size	Uses
Texas A&M* *This bird was originally bred by Texas A&M College in the USA.	18-23 cm (7-9 inches)	A meat bird, slightly larger than Japanese quail. They do lay, but are primarily a meat bird with light (not white) meat and pure white feathers.
Jumbo Japanese	18-20 cm (7-8 inches)	A bird that is both for the table and for egg laying.
Normal Japanese	15-18 cm (6-8 inches)	A great egg layer that can also be used for the table.
Italian	13-15 cm (5-6 inches)	A great layer and table bird.
Spanish	13-15 cm (5-6 inches)	A bird primarily for egg production and is a smaller table bird.
British Range	15-18 cm (6-8 inches)	Very good layers and can be used for the table.
English White	15-18 cm (6-8 inches)	A good layer that can also be used for the table.
Tuxedo	15-18 cm (6-8 inches)	A bird that is a good layer and can also be used as a table bird.
Fawn	15-18 cm (6-8 inches)	A bird that is a good layer and can also be used as a table bird.

A selection of eggs from Jumbo Golden quail

What Do Quail's Eggs Taste Like?

The taste of their eggs is very similar to chicken's egg, but the texture is softer and less rubbery. They are perfect for all recipes in which you would normally use a chicken's egg. When substituting for chicken's eggs in a recipe, the trick is to match weight for weight. For example my supermarket large size eggs weigh around 60g when cracked open. To match this I need around six of my quail's eggs, which weigh around 10g each when cracked open. Of course medium and small chicken's eggs will have a different weight, so you will need to adjust the numbers to the egg size required in your recipe. I would go as far as saying that they taste better in some recipes such as cakes. I've even made quail egg ice cream (which involved the delicate task of separating the yolks from the whites of the tiny eggs – not for the impatient or those in a hurry!). This came about at Christmas, when I wanted to make some mince-pie ice cream and wanted to use larger chicken's eggs for the task, but I had none available. Consequently I resorted to using quails eggs for the recipe. Thankfully the result was a most delicious soft ice cream that all the family raved about for months. The recipe is in the final section if you think you're patient enough to complete the task! On a final note, I would like to add that the small size of a quail's egg can encourage the eating of eggs in those who are normally reluctant to try them. The eggs look delicate and appetising to small children and less daunting than consuming a relatively large chicken's egg.

Eggshell Colouring

The eggs are cream coloured with brown speckles and when you crack them open they are a delicate pale blue inside. There are countless variations in the egg patterning. The variations seem endless.

Endless variations of speckled eggs with no set pattern

Odd Sized Eggs

Although they are a great novelty, finding eggs that are double sized, extra small, or soft eggs formed without a shell, can indicate a stressed bird. It can also happen when a quail first starts to lay her eggs. If you're regularly getting such eggs consider likely stressors and take action to remedy the situation.

Variations in egg size, from left to right: double, average, extra small

A normal sized egg (left) and a large double egg (right)

A selection of eggs showing the huge variety in colour and design

A Tip for Egg Collection

An ice-cube tray makes a great container for collecting your eggs

An ice-cube tray is the perfect size for both collecting and keeping your quail eggs. The eggs fit snugly into the compartments and do not roll around, so reducing breakages. I have yet to find anything better. There are of course special miniature egg boxes to keep your eggs in, made of plastic, but they tend to be more brittle.

Selling Quail Eggs

If you have surplus of your lovely eggs you may wish to sell them on or people may request sales from you. *In the UK the following information applies. It would be advisable to check your own country's legislation before embarking upon egg sales.*

The U.K. is relatively easy to abide with the regulations. DEFRA who are the UK government department responsible for policy and regulations on environmental, food and rural issues, advises that The Eggs and Chicks (England) Regulations 2009 regulating sales of eggs in the shell for human consumption, does *not* apply in respect to quail eggs. There are of course some regulations by which you must abide: quail eggs are covered by hygiene regulations and trading standards, but are free from the red tape that applies to the sale of chicken's eggs. Currently, all you need to do is label them with your address and best before date.

Quail egg cartons are available to purchase at poultry suppliers

CHAPTER 2

QUINTESSENTIALLY QUAIL

Quail are by nature shy and retiring birds and yet also quite curious and totally charming. They will take flight in a vertical manner when surprised and are amazingly good escape artists. They will kick off with their feet and wriggle to escape from your hands. Once confident in your presence, they will take a passing interest in what you are doing and can even be hand tamed, but their natural temperament is to be flighty and shy. If you're looking to buy for your children, please consider that they do not make the kind of pet you can cuddle.

How to Hold Your Quail

Quail are fantastic escape artists. If they escape they are masters of disguise and are hard to find; so learning to hold them is essential. They will struggle using their long legs to kick away from you, flapping their wings vigorously. They have amazing strength in their delicate limbs which are used to propel them vertically when they take flight.

There are a number of ways to hold a quail. One way which is great for beginners as it gives a firm hold, is to place your index and middle finger around its neck and put your hand under its body. This will settle the bird down and it should not move much. If it continues to wriggle you can turn the bird upside down. This disorientates and generally confuses the bird and it will become calmer as it tries to work out where it is.

Never hold your quail by its feet or legs which are delicate and easy to damage. If you do so you risk injuring them or even breaking their limbs. And do not squeeze their rib area too tightly.

Preparing to hold a quail

Supporting the bird with the hand underneath

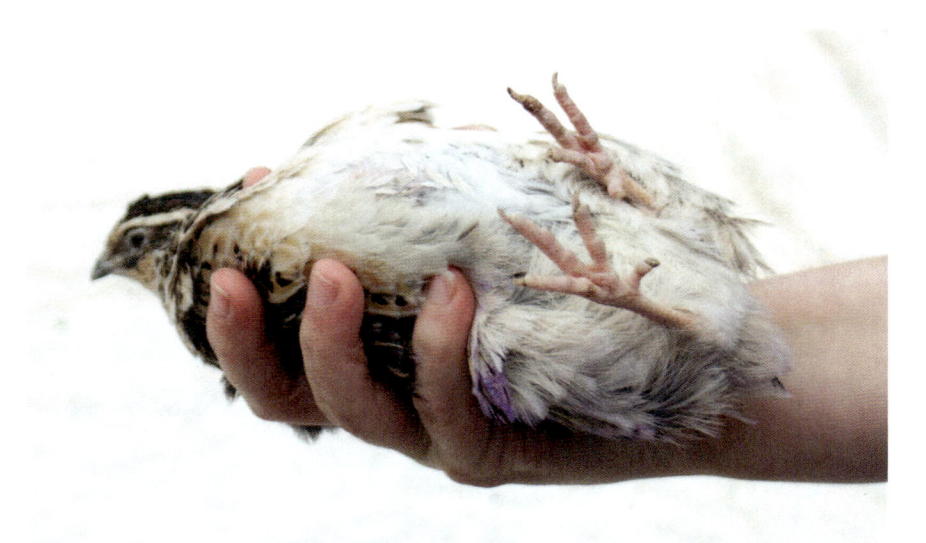

Turning the bird upside down

Taming Quail

You can tame your quail by hand feeding them tasty treats on a daily basis, ideally from an early age. Meal worms are ideal for this as quail find them irresistible. Patience is needed as they are essentially a game bird, not a pet. Talk to them in a soothing calm voice and approach them in a relaxed manner when you offer the treat with a flat open hand. Initially also avoid eye contact with them as they find this threatening. They will soon begin to recognise your voice and be eating out of your hand and competing for the treats. Keeping the quail calm when you approach them will help you do the everyday jobs with ease. Loud noises will unduly upset your birds. Quail may panic when faced with the unfamiliar and don't seem to learn to recognise faces. They do however seem to remember voices and also clothing. Wearing the same or similar clothing when you attend to your birds may induce a more relaxed response towards you.

Hand feeding tamed quail

Surprisingly it's thought that quail may react to different colours. There are countless anecdotes of quail taking a dislike to certain colours, particularly bright ones which may signify danger to them. My own covey panic when I wear the colours peach, red or striped clothing. The latter may possibly instinctively remind them of snakes.

Clothing that flaps around such as scarves can also scare them. So be conscious of what you wear when attending to your quail and if possible wear the same clothing each time when you attend to your birds so that they become accustomed to you as they don't seem to recognise faces.

Will Quail Make a Noise or Disturb My Neighbours?

The song of the female quail seems to blend in with natural bird song and should not bother your neighbours. Male quail will emit a louder song, but once again it is quite a pleasant sound.

Classification of Quail

Looking deeper into the family tree of quail serves to show how modern day quail have been selected for egg and meat production. They are classified as follows:

Class: AVES (Birds)

Order: GALLIFORMES (Game birds and Fowl)

Family: PHASIANIDAE (Pheasants, Partridges and Quail)

Genus: Thirteen different genera commonly split into two categories of 'Old World' or 'New World' quail.

'Old World' quail: *Coturnix, Anurophasis, Perdicula, Ophrysia*

'New World' quail: *Dendrortyx, Oreortyx, Callipepla, Philortyx, Colinus, Odontophorus, Cyrtonyx, Rhynchortyx, Dactylortyx*

The Thirteen Quail Genera

Of the 13 genera, this book is mostly concerned with *Coturnix* 'Old World' quail, so called as they were introduced by European settlers to the Americas. Of these it is the *'Coturnix japonica'* that was selectively bred to form the basis of domesticated quail used today for egg and meat production. Other strains were developed for their ornamental and singing abilities.

**"Wisdom begins with putting the right name on a thing"
(Old Chinese Proverb)**

Having knowledge of the 13 genera is not essential to keeping quail, but it becomes useful when clarifying common names. Sometimes there are numerous local or common names for the same genus which can cause great confusion, but if we use the scientific name for clarification life becomes easier. For example Buttonquail are not a true quail but are from an entirely different family: *Turnicidae*. To add to the confusion 'Button quail' is also the name Americans give to a true quail: *Coturnix coturnix chinensis,* which is also known as Chinese Painted, Asian Blue, King, or Blue Breasted quail. It's when faced with such confusion that the wisdom of giving the 'right name on a thing' becomes apparent.

You can refer Appendix A for a list of the common names and corresponding classification of 'Old World' quail, Appendix B for 'New World' quail and Appendix C for Buttonquail.

CHAPTER 3

QUAIL HOUSING

I've Not Much Space So Where Can I Keep Quail?

Many of us live in houses with small gardens or flats, yet yearn to produce some of our own food. So its great news to discover you can keep quail in a small contained area, unlike chickens. I feel most people will have enough room for these small birds, which can be housed on your balcony, patio, garden or garage.

Hutches

Quail are just perfect to be kept in cages or *adapted* hutches, such as rabbit hutches. I emphasise the word 'adapted' as the environment of a rabbit hutch is too dark for quail to be truly happy. Quail really love the light (plus cover in which to hide). To get sufficient light they need windows on three sides of their housing if you are going to keep them without an outside run.

Runs

If you have the space you can of course go one step further and create an outside run attached to your hutch, or have an aviary. For the purpose of this book I'm primarily considering the home farmer with a limited space and how to keep quail happy and healthy in hutch type conditions. But if you have a garden you may wish to create a run on grass.

If you make an outside run it will need to be moved frequently, so that the quail don't suffer from foot problems. Keeping them on grass will also get them into contact with parasitic worms, so preventative treatment will be needed on a regular basis. You will also need to put them in at night in housing preferably above the ground to deter predators, particularly rats and foxes.

Ideally you should have a solid roof on your quail run (not mesh) for a number of reasons: it provides shelter from the sun and rain and it prevents droppings from wild birds which carry parasites and diseases that can be passed onto your birds.

Can I Free Range Quail?

Quail are game birds and will fly away given half a chance; so they cannot easily be free-ranged like chickens, geese and the like. They also do not tend to take shelter when it rains, nor do they have the instinct to return to their run at the end of the day, making them difficult for free ranging. An outdoor run attached to their housing, complete with a roof to protect your birds from the worst of the weather, makes a good compromise.

A Golden quail enjoying an outdoor run attached to its shelter

Basic Housing Requirements

For enjoyable and easy quail-keeping which equates to lots of lovely eggs without too much effort, you will need to consider some essentials that will ensure stress free and happy birds. It really is worth getting the basics right the first time. Get them wrong and you may encounter feather pecking, bullying birds, cannibalism, low egg production and many other problems.

Six Elements of Great Quail Housing

1. Size of Accommodation

Overcrowded birds will begin to peck each other and be quite unhappy. From trial and error I found that the minimum space required in a hutch type of environment with no outside run should be a minimum of 1½ square foot per bird (0.14 m²). This is a comfortable enough space provided that additional hiding cover for them is provided. Any less space than this and you will begin to encounter stress related problems.

Any more space than this is a bonus. So in a hutch 5 ft by 2 ft (1.55 m by 0.6 m) you will have ten square feet (0.92 m²) and therefore seven birds should be comfortable and happy. I know that intensively farmed birds have very much less space than this. Indeed some have suggested space of just 4 ¼ inches by 4 ¼ inches (11 cm by 11 cm) per bird! Just room enough to stand. Under these conditions which are the equivalent of battery farming, they are merely surviving and not thriving. And I think that nowadays we are more concerned with animal welfare. Personally I would like to create an environment where birds not only live and produce successfully, but also have a happier less stressed life.

2. Avoid Damp

Damp conditions create an ideal environment for many diseases to fester and quail are unhappy living in damp or wet quarters. With regard to outside runs you would think that being a game bird quail would not mind getting wet, however as previously mentioned this is not the case. It's important to note that if it rains most quail will not usually take shelter, but just stand in the pouring rain getting cold and wet and will need to be rescued and made warm and dry if this happens. This comes as a surprise and is hard to comprehend; perhaps taking shelter has been accidentally bred out of them over the years. So if you create an outside run, ensure that it's sufficiently covered on all sides to prevent them getting wet.

3. Avoid Windy Conditions

Even the most even natured quail dislikes windy conditions. The most gentle of breezes have the potential to drive them to distraction. So they need protection. Putting the hutch into a shed will help or glazing the wire mesh windows with Perspex will cut out draughts. Although do ensure there is sufficient ventilation to avoid an accumulation of ammonia from their faeces, which can damage their tiny lungs and create respiratory problems.

4. The Perfect Temperature

The ideal temperature for adult quail is 16°C (61°F) to 23°C (73°F) and a relative humidity of 30-80%. If you don't keep your quail in a shed or garage and you decide to keep your quail outside you can improve their conditions by taking the following measures:

Place a covey of around 20 birds together so they can huddle to keep warm (you need to get your hutch size correct to accommodate this number of birds).

Place some sort of covering around the outside to windproof the front, but ensure there is still sufficient ventilation. I have used a translucent shower curtain with a degree of success that allows in light, but blocks out the wind. I've also used acrylic windows to block out the wind (ensuring there was still enough ventilation).

Provide some heating. For example have a qualified electrician put in a low wattage heat source such as a tubular heater covered with air bricks, so the birds don't burn themselves on it. Precautions to take also include needing to have good air flow from the heater to the bird area and avoiding condensation and a thermostat or thermal cut-out, so that the birds are not over heated.

5. Perfect Lighting

In a natural environment your birds would go off lay for three months a year. Some people like to give their birds a rest and say it prolongs their lives. Others continue to provide light through winter by artificial means. The choice is yours. Quail require 14-18 hours of daylight per day to encourage maximum egg production. So if you want them to lay over the dark winter months you will need to provide extra artificial lighting from autumn to spring to keep those eggs flowing.

The intensity does not need to be very strong; indeed you can use low level lighting. I have managed to maintain very excellent egg production in the dark month of December with nothing more than a battery powered L.E.D. reading light placed directly in front of each of the cages. A low wattage bulb will do if you've got an electricity supply. In a small shed a 25 watt bulb is adequate for about a dozen birds and a 40 watt bulb for around 50 birds.

If you have a small garage sized building, then a fluorescent strip light would be sufficient. You could also add a timer to your lights so you don't have to run around switching them on and off.

Also consider the heat emitted from light bulbs and ensure they don't create an excessively warm environment for the birds not to mention avoiding fire hazards from hot bulbs.

Finally be aware that very bright lighting when given to all male groups will increase their desire to fight. Therefore it is best avoided in preference to more subdued, less direct lighting.

6. Location

Quail love the sun and enjoy an eastern facing home. Providing them with eastern facing housing especially in colder climes ensures they will get the morning sun and so heat up quickly on cold winter mornings.

Hutch Designs

Of all your decisions your choice of housing will have most bearing upon your quail-keeping experiences, both good and bad. So it's worth putting in the time and effort to get it right.

Heated housing with windows on three sides and additional lighting

Height

A height of 12-18 inches (30- 45 cm) should prevent injuries owing to a quail's rapid vertical take off when startled. The ceiling is best made low to avoid injury. Some keepers even place soft padding on the ceiling. By their nature they take off vertically at speed when surprised. So if your roof is too high they will hit their heads. If you wish to have a high ceiling, try to make it really high such as the height of an aviary.

Legs

Your quail's accommodation is best placed upon legs, so creating a space under the hutch to deter predators, which can occur even in an urban environment. Frequent pests are rats, cats and foxes. Rats in particular just love to eat a tasty quail snack. However they dislike going out into the open as it makes them vulnerable to attack; so an empty space beneath your hutch is a great preventative measure. Rats will also be attracted to any food spillage, such as grain. Therefore keep the area around your hutch spotlessly clean to remain rodent free.

Single or Double Layer Housing?

There are many impressive double layered style of housing available to buy, but quail are ground loving birds, so most of the time they will only live on the lower level with only one or two venturing upwards to the higher level. Even if you provide a walk way up to the next level they seem to dislike walking up a steep walk way. Consequently I would suggest only creating single layer accommodation.

Flooring

The ideal housing should have the attribute of being easy to clean as no body wants to spend hours doing this task. Ideally it should have a slide out removable floor for the ultimate in speed cleaning. The surface should be smooth and free from ridges, so that it makes the task easy. My first quail covey lived in an adapted rabbit hutch that had a tongue and groove surface that took forever to clean and this type is not to be recommended as it can harbour mites and disease. I resorted to putting vinyl over the surface (under the wood shavings) to make it easier to clean.

Some people opt to keep their birds on wire flooring. You can start your chicks on wire from six days old using quarter inch (6 mm) welded mesh and then move them onto one inch by half inch (25 mm by 13 mm) welded mesh from 14 days old. Remember to use the welded sort as other types will cut a quail's feet. Some people claim the advantage of keeping quail on wire is that it's healthier for the birds. You've no bedding material to dispose of, only faeces. These drop through the wire onto a collection tray. The rate of cleaning remains the same but the smell can be worse as it is not covered by any bedding. The disadvantage of keeping your birds on wire is that the quail's natural behaviour is inhibited. For example they cannot enjoy scratching around in their bedding.

I feel strongly that if you do choose the wire mesh option please consider providing natural features such as a sand bath and a non-wired area where the quail can rest their feet and some natural plant cover in which they may hide. It's a pretty bleak prospect for a quail to live out its life in a sterile battery-cage type of environment without any natural materials around and unable to exhibit normal behaviour.

Windows

The wire over the windows is best made of smooth welded type mesh. If not should birds fly at it, they will cut their beaks or wings on it and injure themselves. It only takes a small event for this to happen, such as a stranger walking by and some of your birds may simply take fright. I use 1cm square welded wire mesh on the windows which is small enough to prevent adult birds from putting their heads through the holes. If where you live wire is available only in inches, then half-inch welded wire should be suitable.

Welded wire mesh (left) and twisted wire mesh (right)

Quail on welded wire mesh flooring with sand bath in background

Seven Essential Items for Inside the Quail Hutch

The following items help to make a cosy covey of quail:

1. Absorbent Bedding

Wood shavings are preferable to hay or other material, as it is easier to clean and unlike hay it is not prone to having mites in it. If you use wood shavings dust extracted is preferable. This can work out quite expensive unless you source large bags of it. You can offer some of your eggs in exchange for it of course!

2. Small Sized Grit

In the absence of teeth, quail have a gizzard which is responsible for grinding up their meals and grit plays an essential part assisting with the process. Offering grit may seem pointless until you learn that without it your quail will die at a younger age, possibly by choking if not from lack of nutrients. Chicken grit is too large for quail, so ensure you provide small particle grit which you can get from pet suppliers.

3. Food Dishes

Quail love to scatter their feed as they eat and given the opportunity they just love to jump into it kicking it around wasting lots of grain. Also if it's left uncovered the birds will poop in their food increasing the chance of illness and spreading any infections. I would therefore recommend the use of covered food dishes. Generally they have cut-out holes in the container where the birds can place their heads to eat but are unable to put their feet and bodies into it. Some like the green food container below are gravity feeders and can dispense grain as it's eaten.

Covered feeders

Covered drinkers

4. Water Dishes

Clean water is needed throughout the day, not just once or twice per day. The birds must have freely available clean water when ever they want it. Often this is overlooked, but is a most essential element of their care to prevent against diseases and promote good health. Preferably the containers should be of the covered type to prevent water contamination as they are prone to pooping in it at any given opportunity.

5. Dried Cuttlefish Bone

Quail love trimming their beaks on the Cuttlefish bone. It's given for beak condition and extra nutrients and can be obtained cheaply from pet suppliers. You can fasten this to the wire on the cage.

6. Hiding Places for the Birds

The more places quail have to hide, the more relaxed they will feel. Paradoxically you will actually see more of them because of this. It's simple to create hiding places using everyday objects. For example you can make a small bush structure using clippings from evergreen shrubs joined together with string and placed into a pot. Or you can create a series of small or larger hides using a collection of shoe boxes. The quail will love to hide in these structures and take peeks at the world while feeling safe and secure. These are just two suggestions for the type of structure you can make, but the only limit is your imagination.

7. Sand Bath

Quail just adore a sand bath. It is relaxing for them and reduces stress and pecking problems and it's wonderful to watch them having such a great time. Find a deep enough container whereby the birds can jump in and have a good scratch around, without the sand spreading around covering the rest of the cage. It is best located in an area of the covey that gets the morning sun, so the sand can warm. The sand will need sifting and cleaning regularly. It's a great place in which to put anti-mite powder, so that they are frequently medicating themselves.

Quail enjoying a tranquil moment in the sand bath

A popular place to unwind for both cocks and hens

A cock dusting his feathers in the sand bath

CHAPTER 4

QUAIL FOOD

Quail need the correct nutrients to thrive. When I first considered keeping quail I read a few magazine articles and online forum posts where quail owners stated that they fed their quail chick crumb (designed with chickens in mind) with great success. So I went ahead and did the same, but after about five months I noticed some problems: their feathering condition was poor and some were laying eggs with soft shells. As a result I compared the complete nutritional requirement of quail to the chick crumb I was using and to my horror found that my chick crumb was most unsuitable for them as the nutrients were sub-optimal. When I searched around so was every other chick crumb that I found on the market at the time. Unwittingly I had been making my birds sick with inappropriate feed.

Recommended Nutritional Requirements of Quail

Age and Type	Protein %	Calcium %	Phosphorus %	Methionine %
Chick for egg production (0-6 weeks)	24%	0.85%	0.60%	0.50%
Chick for meat production (0-6 weeks)	24%	0.85%	0.60%	0.50%
Adult layer from 6 Weeks	20%	2.75%	0.65%	0.45%
Adult meat bird from 6 weeks	18%	0.65%	0.50%	0.40%

Recommended Nutritional Requirements of quail

Quail feeds with the complete and correct nutrients are not easily sourced. I have found that just because it says '*Quail Food*' on the label does not always mean it contains the correct nutrients. Many use turkey feed for their quail, but again the nutrients are frequently insufficient. The advice I would give is to check the nutrient label on the bag and phone the company for more information if it's not listed.

I have since managed to source a relatively cheap organic food that contains all the correct nutrients. Before I changed their food I was concerned that overall my flock's weight was too low and their feathering was poor, they looked and felt too thin around the breast bone and seemed generally under par. I wanted to know if their new food would make a difference to their condition, so I weighed two of the birds: a standard quail and a 'Jumbo' quail. After four months on the new improved feeding regime that contained all the correct nutrients I weighed them again. While it is a very small sample and a one-off trial the improvement noted was impressive.

Improve Your Bird's Health with Correct Nutrients

Chart Showing Weight Gain of Two Quail

Type of bird	Weight at start	Weight after four months	Gain / loss
Standard quail	235 g	269 g	35 g gain
Jumbo quail	370 g	420 g	50 g gain

Feeding the birds with food that contained all the correct nutrients made a wonderful difference to their health. After four months the condition of all my flock had greatly improved; their feathers were lush and plenty. The 'standard' quail which had been the worst specimen was now bright and healthy. Her breast bone was no longer prominent and she had gained weight. The Jumbo quail had also gained an impressive 50 g in weight. People will be quick to tell you that a fat bird does not lay eggs and being fat can cause problems. However these birds had been in poor condition owing to the poor feed I had been administrating and I believe the new feed was at the heart of the impressive improvements.

The Role of Nutrients

Methionine - One of its uses is for feathering. Birds deficient in this may eat their own feathers.

Calcium - The birds use this for bone formation, eggshell production and blood clotting. Too little calcium can cause soft eggshells, demineralisation of the bone and fracturing. For additional calcium you can add oyster grit (small size), or purchase calcium carbonate (limestone) from a feed merchants who may advise you how much to add to your current feed, or you can purchase it in liquid form to add to their water. The amount will vary depending upon how much calcium is already in the food.

Phosphorus - This is used for egg formation, bone formation, metabolism of fat and carbohydrates.

Protein - Proteins are essential for life. The bird uses this for growth and development.

How Much Food Do Quail Need & What is the Cost?

Allow your birds around 45 g per day for Jumbo quail and less for smaller birds. Having this knowledge you can calculate how much your birds will cost you. I've also read that several accounts of quail consuming much less at around 130 g per week. As mine eat much more at 315 g per week I can only assume my calculations also include some natural wastage such as the dusty crumb I will discard as the feeders get empty; spoiled food (with faeces) that I remove and also that other people may have formulated their calculations from data on smaller birds. One of the largest commercial farms in the UK feed their Jumbo birds an allowance of 50 g per day, so I think my calculation of around 45 g per bird is on track.

As the seasons progress you may notice the amount the birds eat varies owing to different weather conditions. They eat more when it's cold and less when it's hot. Sometimes they seem to eat less for a couple of days for no reason that I can yet perceive. Then they will make up for it in the days later

Favourite Snacks and Treats

Additional snacks can help with overall health and well-being and also reduce pecking. Additional greenery will make the yolks of their eggs more of a rich yellow.

In addition to their main diet there are a number of additional foods or nutritional snacks that you can give your birds.

Salad greens - Lettuce (many quail love this), spinach and chicory.

Garden weeds - Chickweed and dandelion.

Mealworms - Dried or fresh, allow a maximum of five per bird as they have quite a high fat content.

Not all quail like the same foods. The best advice I was given is that the quail will let you know what they like. Offer them a new food a couple of times for them to get used to seeing it. They will have a chance to examine it and peck at it. If they eat it, you will know they like it. If they don't eat it, remove it the next day and try again another time. After a couple of times they may try it and enjoy it. For example sometimes they can prefer foods that are in season, so will eat apples in autumn, but not spring.

Foods to Avoid

A word of caution: toxins can take time to build-up, so be cautious and check out the suitability of the food first, before feeding it to your quail. Many harmful foods will not instantaneously kill your birds, but take time to build-up in their bodies. Harmful foods include:

Raw green potato peelings - These have a toxic substance called solanine, which is a glycoalkaloid poison found in species of the nightshade family, such as the potato. It's found in the green sections of the plant such as leaves, stems and shoots and any tubers that have turned green will include it.

Salty Foods - As salt poisons their small bodies.

Dried or undercooked beans - As they contain hemaglutin which is poisonous to some birds.

Avocado skin and stone - As they contain low levels of toxins.

Chocolate or foods with sugar - Chocolate is bad for their systems and is poisonous to most pets.

Henbane - A poisonous weed.

CHAPTER 5

QUAIL CARE

Before beginning to keep quail it's important to appreciate the amount of time and work you will be doing. They do not take a large amount of time or effort but there are regular jobs you must undertake to maintain a healthy covey. Routine jobs are all part of any poultry care and if you schedule them into your daily routine they should not be too arduous.

Daily Jobs

- Feed and add clean water.
- Check levels of grit.
- Add small green treats.
- Make a quick visual check of their health.
- Collect any eggs.
- Check the temperature of their accommodation.

Weekly

- Handle all of the birds; give a general health check of toes, eyes, beaks and feather condition.
- Clean out their accommodation.
- Clean the feeding and watering equipment and disinfect.
- Sift the sand bath to remove any grime and add anti-mite treatment and replenish with more sand. This task may need to be done more than once a week.

Monthly

- Worm your birds if they are kept on soil.
- Check the structure of your accommodation and do any repairs.

Six Monthly

- Give an extra thorough cleaning finishing with a Red Mite treatment.
- Add/Remove any Perspex window covers depending upon the time of year.
- Consider replacing any birds.

Cleaning the Housing

Making time for cleaning your housing is very important as the key to good health for all poultry is cleanliness. Cleaning out must be done a minimum of once a week to maintain the health of your flock, prevent a build-up of faeces on the bird's feet and to prevent mite infestation. It will also keep flies and bad smells at bay and is an easier task if done regularly.

To clean your housing you will need: absorbent bedding (preferably not straw as mites are often hiding in it), a scraper tool to remove any dried faeces makes quick work of the job ahead and mite control spray or powder.

Method:

1. First remove the quail.
2. Next remove the old bedding material which you can compost. It makes an excellent fertilizer.
3. Apply anti-mite control.
4. Clean all water and food containers.
5. Put in new bedding.
6. Add Red Mite control following instructions carefully.
7. Clean the sand bath (sieve it).
8. At this point do a health check on your birds, checking their feet, beaks, feathers and signs of pecking from other birds.
9. Return your birds to their bright clean home.

Although it's not usually a favourite chore, it will ensure your birds are healthy and therefore your eggs will be too.

CHAPTER 6

PECKING PROBLEMS RESOLVED

Bullying and pecking also known as cannibalism is unfortunately a common issue. The behaviour includes pecking of the eyes, top of the head or beak and feather plucking from any area of the body. It can be severe, with birds drawing blood and even killing each other. It can happen quickly so once you spot a problem act immediately to prevent further damage.

Above all it's important to note that *any form of stress* will result in a pecking problem; so the first step is to remove any stress factors in your bird's environment.

Factors Causing Pecking

Many things can cause pecking including the following:

- Overcrowded housing.
- Territorial disputes.
- Introduction of new stock.
- Young male birds reaching maturity.
- Mite/Lice infestation or other health issues.
- Lack of stimulation or boredom.
- Damp or moist housing or bedding.
- Draughty accommodation.
- Windy conditions.
- Adverse temperatures, either too hot or too cold.
- Weather changes affecting their housing such as rain or frost.

Reducing Pecking in Quail

To avoid your birds becoming aggressive, try the following suggestions:

Reduce overcrowding - Generally if you keep your quail in a hutch type of environment, try to give them 1.5 square feet per bird as a minimum.

Keep their environment interesting - Place in their housing things to peck or jump on, for example small blocks of wood.

Add screened areas in which they can hide - For example by making wigwam structures from evergreens, or twigs in which the birds can hide.

Create a great hide - For example using upturned boxes with a door cut into them to create a small house. Your quail may both hide or lay eggs inside there.

Avoid very bright lights in all male groups - This will reduce aggression.

Add twigs or greenery - You can add these to the front of the hutch so they can conceal themselves or place them inside. The more you offer them places to hide the more you will see them.

Move food and water dishes - As the birds are territorial you can move around the food and water. This reduces tension.

Make a sand bath in which they can relax - You can add anti-mite treatment so they are self-medicating and prevent problems from arising.

Give them green vegetables to peck - Raw vegetables or clean weeds from the garden (that have not been chemically treated or in contact with dog or cat excrement) will keep their minds busy.

Provide areas for them to climb on - Wooden blocks or boxes for example so that the bullied birds can escape trouble by being higher and out of the way.

Keep the hutch within the correct temperature range - Too high or low will cause your birds to be stressed. Keep them between the ideal temperature range of 16°C (61°F) to 23°C (73°F).

Make their home draught free - Quail find draughts and wind disturbing.

Add a source of heat - To keep the temperature in the correct zone.

Put the hutch in a shed or garage extra warmth - Quail enjoy the shelter in the winter months.

Provide a dried Cuttlefish bone - Position it firmly against the wire mesh, much as you would do for a pet budgie, so they have something to peck at.

Keep a male bird with your females - Some believe this reduces bullying in an all female environment. However you do not need male birds to increase or stimulate egg production.

Do not keep a number of mature male birds together - They will fight to the death.

A bullied quail its pecked feathers appearing moth-eaten

CHAPTER 7

HEALTH

It can be difficult to interpret the signs of sickness in quail as they are quite hardy and will only begin to reveal the extent of their illness when they are very poorly indeed.

Some signs of a sick quail include:

- Withdrawal into a quiet part of the cage.
- Droopy feathers.
- Ruffled feathers.
- Humped up posture.
- Not eating or drinking.
- Diarrhoea.

A Hospital Cage

When they do become sick they need to be placed away from the rest of the flock, for peace, quiet and recuperation. A separate hospital cage will allow you to provide the best environment for recovery. If you find any of your quail showing signs of sickness, they should be taken out of the flock immediately and put into a hospital cage.

Such a cage needn't be anything fancy or complicated. It could be a makeshift area such as a cardboard box or spare cage if you have one, with shavings inside of it. Always provide water and food, although poorly birds tend to drink more than eat. Place the cage in a warm and draught free spot, with a gentle heat source such as a light bulb. Alternatively a short period of an infra red heat lamp will assist recovery. Make sure that they have a cooler area they can get to if needed. If they are unable to move do not expose them to too much heat.

Birds that are listless can invite pecking from others and isolating them helps them have a quiet place to recover and helps to prevent the spread of infection. Placing your sick quail in a cosy warm hospital cage can make the difference between life and death when all other measures have been taken. Time out can help badly pecked birds to rest and recover; although you must also remove the culprit to break the pecking habit.

Beak and Toenail Health

Sometimes toenails and even beaks grow too long and need to be trimmed. For new quail keepers this can sound daunting and potentially painful for the quail. However done correctly it should be painless and beneficial. It should be undertaken only if required.

Why Trim the Beak?

Trimming a bird's beak may become necessary if it grows too long and prevents easy feeding. If it's left untrimmed the bird will become thin and malnourished. The breastbone will protrude and general condition will deteriorate.

Trimming the beak is not to be mistaken with heat cauterising the beak as done with battery chickens and other game birds kept in captivity; but it is merely like trimming your toenails and should be painless, if done correctly. When required I use nail clippers and trim the upper beak only. This action should be painless. I make sure that like finger nails that I don't clip too much off and can always go back to remove more if necessary.

An over-grown beak

Some people trim beaks under the assumption that doing so may reduce the damage a bully bird does to the others. The logic being that a blunt pointed beak (made blunt by something like nail cutters), would do less damage than a very sharp one. However they quickly file their beaks back to a point within a few days, consequently I feel it's a waste of time. A better solution is to reduce their overall stress and consequently reduce bullying, without the time and effort involved in catching your birds and potentially upsetting them and even causing a reduction in egg laying.

How to Trim a Beak

It's not a difficult task, although daunting at first. However it should be painless for the bird. So I've listed the procedure below.

You will need: a small pair of sharp nail cutters (use safety clippers, not nail scissors. See the following photographs if you are unsure)

Method

1. Hold your bird securely in your non-dominant hand.
2. Hold your clippers in your dominant hand.
3. Hold the bird up to the light and locate the darker area/vein at the top of the beak.
4. Gently prise open the beak.
5. Trim the upper beak, but not as far as the darker area/visible vein. Do not trim the lower beak.

If you do accidentally clip to far up the beak and cause bleeding, it should stop quite quickly. Just take measures to separate the quail from the others as they get attracted to blood and will peck the bird. Then once it's stopped bleeding return it to the other birds, having wiped off any excess blood if necessary. Her beak will be tender for a few days so ensure her food supply is increased to the point where she will not have to peck at the bottom of a food tray.

The bird's beak will grow back relatively rapidly and may contribute to unwanted weight loss as the bird is less able to eat. So trim it whenever it gets too long. The bird will then be able to feed easily. She will shape her beak on Cuttlefish bone if available, to make it sharp once again.

Nail clippers at the ready preparing to trim the beak

The bird's beak trimmed to normal length

Why Trim Your Birds Toenails?

If you're keeping your birds in a hutch, on either sawdust or on wire, they won't get enough abrasion wear down their nails and they will grow too long. This may exacerbate toeballing in the short term and in the long term the feet may become deformed and the nails to drop off, all of which can be very painful. When you do your weekly clean out inspect your birds' feet and trim their nails when needed. It's painless for the bird and just like cutting your own nails, make sure you don't cut to the quick.

How to Trim Your Quails' Toenails

You will need: a small pair of sharp nail cutters (once again use the safety clippers rather than scissors).

Method:

1. Hold your bird securely in your non-dominant hand.
2. Hold your clippers in your dominant hand.
3. Gently trim back their nails. Trim a small bit at a time. You can always trim more.

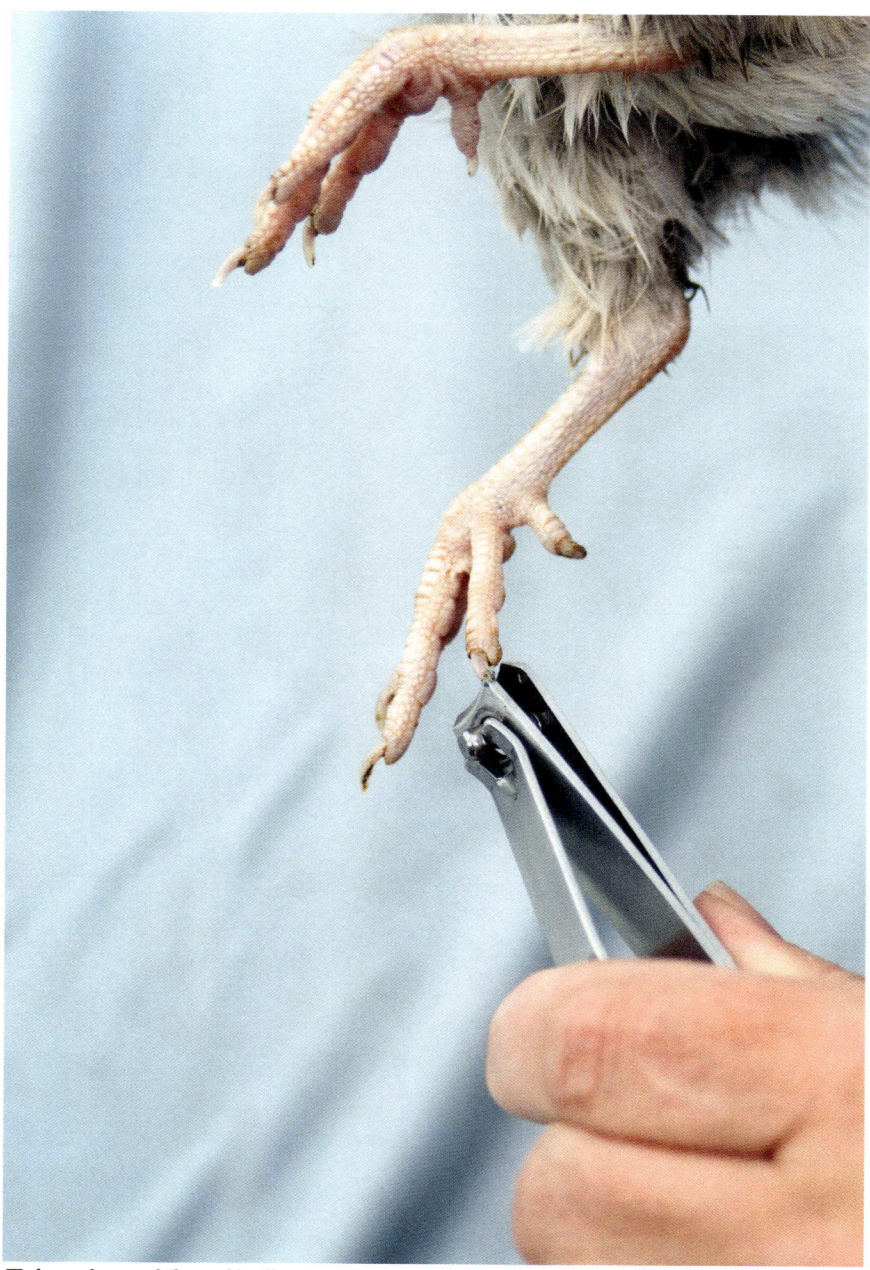

Trimming with nail clippers

Prolapsed Vent

Sometimes quail will suffer a prolapsed vent area, which looks like a swollen red lump around the size of a largish marble protruding from her behind. This condition is caused when a bird lays an egg a bit bigger than normal and her vent (cloaca) turns partially inside out. Over-weight birds are also prone to getting this condition.

A prolapsed vent seen as a red lump protruding from the quails vent

How to Treat a Prolapse

It is important to separate her from the other birds as they will be tempted to peck at the area and doing so will kill her.

Hopefully you will discover the bird's condition before the prolapse has dried out significantly. If it has dried out, moisten the area. It's a good idea to very gently rinse the area as you are going to re-insert it and you want to minimise infection. At this point check that the area has not been pecked by the other birds.

Unless you hold the bird correctly she will struggle. So turn the bird upside-down in the palm of your hand with your index and middle fingers gently around her neck. She will go into a relaxed state in this position.

Next wash your free hand thoroughly so that you don't introduce any extra germs. Then very gently apply pressure to the centre of the prolapsed vent and insert your little finger into the cavity. The swollen prolapsed vent will return whence it came! To do this with your own hand is not as vile as it sounds and just feels warm. Neither is it particularly filthy job, but do make sure you wash your hands afterwards!

Hold your finger in the cavity for about 30 seconds. You may then feel the bird's muscles contracting around your finger. At this point very gently remove your finger and the prolapsed vent should hold back inside their body cavity. After you remove your finger you can also apply very gentle pressure on the outside of the vent for around 20 seconds, just to stop it popping out again.

The bird now needs to be kept calm and quiet to assist with healing. Some people advocate putting the bird in a dark quiet place on its own for an hour or two, but without heat in case the prolapse re-occurs. You want to prevent the prolapse from drying out.

To guard against infection you may like to add soluble antibiotics to her drinking water for four to five days.

Once a bird has had a prolapse, it is likely to re-occur. You may need to repeat this as many as 12 times, for up to a week. Initially keep checking every half hour or so until you're confident it's going to stay put. If it does not stay put, repeat the process for up to a week. If by then it's not worked, you will have to cull the bird.

Toeballing

Quail can suffer from the debilitating condition known as toeballing. The condition as the name implies is a hardened ball of material attached to the foot. It can occur when they stand on mud, their own droppings, damp material, or because of walking on dirty bedding material such as wood shavings. If the latter becomes a problem, then it's necessary to decrease the length of time between bedding changes.

Imagine walking across a field of mud, the longer you walk the greater the amount of mud sticks to your shoes, until eventually your feet are nothing but a big blob of heavy mud. If you then sit down in the sun to rest and the mud dries, you can imagine how hard it is to get the now hardened mud off your feet. I'm sure you get the picture. In the case of quail, the material dries into hardened lumps around the bird's toes and walking is hindered. They cannot remove the lump themselves. This condition can develop over just a couple of days, so it's important to do a quick visual check of the bird's feet on a regular basis.

Large Toeballs

To remove the material, you soak their feet in a container holding tepid water, a good amount of washing up liquid and a small amount of Epsom salts. You can also put in wood shavings to stop them slipping and to provide abrasion. Place the bird in a confined area to soak it for half an hour. After soaking thoroughly you can carefully pick off the toeball, being extra cautious not to accidentally remove the toenail. You can use running water if necessary, to further loosen the material. It may take more than one attempt to remove it, or be done over a couple of days. Having someone to help is also useful.

Bumblefoot

Bumblefoot is a septic abscess caused by a small wound to the foot. The wound may be as a result of walking upon rough surfaces, for example concrete. Their delicate skin is pierced by the rough surface and infection sets in. A hard lump forms beneath the foot and usually the quail will be limping. If the infection is rife, the foot will also be hot and swollen.

The infection can be treated a few different ways depending upon your vet's advice. Usually one of the following methods is currently employed.

- Lance the foot, clearing out the puss, followed by wound spray or footbaths of iodine.
- Antibiotics followed by draining of the wound if the former does not work.

If you've got one bird with the condition, it's easy for it to spread through your flock, because the condition is caused by the staphylococcus bacteria, which can easily get into the wound of another birds foot. It's also worth noting that humans can be affected, so wear gloves and dispose of any dressings carefully.

You can tell the difference between this condition and other foot problems by the hardness of the lump. If it's a hard lump then it could be Bumblefoot.

Red Mites (*Dermanyssus gallinae*)

You will be a very lucky poultry owner if your birds never get afflicted by these vampires of the poultry world. They are almost invisible to the naked eye and hide in the crevices of poultry houses. They come out at night and suck on the blood of sleeping birds, usually in the breast and leg regions. Egg production can be affected, and the birds can become lacklustre and thin, even anaemic. In young birds anaemia can be fatal.

Additionally, an infestation can cause your birds to be agitated and they may peck at each other, so prevention is better than cure.

To obliterate Red Mite, firstly treat the housing with anti-mite spray when cleaning out. Then put a treatment on both the bedding and also the birds on a regular basis. One popular treatment is Diatomaceous Earth, also known as Diatom Powder. Because it's an inert mineral it has the advantage of no egg withdrawal or chemical residue. It's a naturally occurring sedimentary rock, composed of 80 to 90% silica (when dried out). It's an abrasive feel to it and works by absorbing the fats from the waxy outer shell or exoskeleton layer of the mites. They then dehydrate and die when they come into contact with it. You can add it to bedding and directly onto the birds. When you put powder on the birds, place it under their wings and around their neck and rump. You can also treat their sand bath with a good dose, for automatic dusting of the birds.

Scarily, Red Mite can survive for up to 10 months in empty housing and like most of us they just love to move into brand new and clean accommodation. Therefore be sure to be vigilant regarding this nasty pest. Just because your poultry house may be quite new, does not mean your covey are exempt from this problem.

Scaly Leg Mite (*Knemidocoptes mutans*)

A condition known as Scaly Leg is caused by the Scaly Leg Mite *(Knemidocoptes mutans).* These mites burrow into the scales of the legs, causing a thickening of the scales so giving the impression that they are protruding outwards, rather than lying flat and smooth. It can result in the legs looking white and crusty. This particular type of mite spends its entire life-cycle on the bird and is spread by contact between the birds.

To treat it, you can buy a variety of specifically prepared remedies from poultry supplies shop. Do not be tempted to pick off the crusts from the legs as doing so will cause damage to the bird.

Lice

Lice are parasitic and if present on a bird can be found on the body, under the wing and around the neck. Their eggs show as a white build-up at the base of the feathers around the rump. They irritate the birds and cause them to become restless. Body weight and egg production may drop. As the lice begin to irritate the birds, they may damage their own feathers by pecking or scratching areas. The rest of the covey may join in pecking the infested bird.

If you have been treating your birds with Diatom Earth (see the treatment for Red Mite), then your birds should also be protected for lice. The same treatment dehydrates these little beasties, who can multiply at an alarming rate. It takes four to seven days for the lice eggs to hatch. The length of time between lice eggs hatching and becoming adult lice is around 21 days. They mate on the bird and egg laying begins two to three days later. As each female louse lays from 50 to 300 eggs; the problem can quickly multiply unless checked.

A Comparison to Distinguish Between Lice and Mites

	Lice	Mites
Size	2-3 millimetres long	1 millimetre diameter
Speed	Fast-moving	Slow-moving
Colour	Straw-coloured (light brown)	Dark reddish black
Egg location	Base of feather shaft	Along feather shaft
Egg colour	White	White or off-white
Best detection time	Daytime	Night time of Daytime
Location	Lives only on host	Lives on host and in environment

Worms

If your quail have an outside run on the soil parasitic worms can be picked up from the ground. Caged quails are unlikely to have these. Intestinal worms cause weight loss and poor feather condition. Other signs may be an increased appetite, a drop in egg production and possibly loose droppings. If your quail are kept on soil you need to worm them as a matter of course, using a proprietary wormer.

One excellent natural preventative (not cure) is Apple Cider Vinegar added to their drinking water. You can find information on how to make it within the pages of this book.

Coccidiosis

Coccidiosis is a microscopic parasite that lives in the intestine of the host quail. They are spread via the faeces, which can contaminate your bird's food. They can pick up the infection from contaminated areas such as the soil, accommodation, food and water troughs, which may have been contaminated by other infected birds or birds that carry the disease.

One single oocyst (microscopic egg) will multiply by many thousands after passing through only one bird. They pass out into the faeces, where they contaminate food and water sources. Quail are great at pooping in their food and don't seem worried about doing it in there. Consequently vigilant cleanliness and raising the food and water dishes up slightly can prevent transferring the disease. The infection just loves a moist environment and dies quickly in dry litter. Therefore once again, cleanliness is paramount in preventing spread of diseases.

Coccidiosis can be suspected in quail that have diarrhoea that includes bloody, white, greenish and slimy diarrhoea, ruffled feathers, soiled vent feathers, general malaise, lack of appetite, weight loss, protruding breastbone and dehydration.

At autopsy coccidiosis can be seen as a spotty and inflamed intestine.

Adult quail are frequently carriers of the disease, without themselves being infected. Yet they can then infect other birds including your young chicks, for which the condition may be fatal. It shows mostly at ages three to five weeks.

To prevent outbreaks take the following measures: good sanitation and litter management, prevention of wet or damp spots in their bedding and the addition of Apple Cider Vinegar to their drinking water (see below). Also, some starter food for chicks includes a coccidiostat. This does not mean they won't ever get the problem; it was developed for use in good clean environment, so it is just another preventative measure. Medication can be obtained from a vet and added to their drinking water or food.

Depending upon the drug prescribed; follow the withdrawal period (of egg consumption) carefully.

Ulcerative Enteritis (Quail Disease)

Ulcerative enteritis also known as Quail Disease is a bacterial infection causing ulcers in the small intestine of the bird. As a result it cannot absorb nutrients from its food and becomes extremely thin their breast bones becoming prominent.

Enteritis in quail results in greenish diarrhoea that differs from coccidiosis in that it is less slimy. The bird can look depressed, humped up with ruffled feathers and not wanting to walk around. Death happens over a period of days and can wipe-out almost all of a young flock. Treatment is by antibiotics.

The disease is spread from bird to bird in their faeces, so it can also be spread by flies and other insects and by carrier birds (those that have had the disease, but recovered). Good management practices and cleanliness will help prevent outbreaks, including:

- Keep water and feeders clean from faecal matter.
- When breeding quail keep visitors away from the area.
- Hatch your own chicks to increase your flock, rather than purchasing and so possibly introducing disease into your flock.
- Have good insect and rodent repellents.
- Keep your hands and clothing clean when dealing with your quail chicks.

Cuts and Abrasions

Quail are attracted to the colour red and will curiously peck at it. So if one of your quail has a cut or abrasion or is pecked until it bleeds any accidental damage will quickly become worse or even fatal as the flock investigates it.

It may be necessary to remove the pecked bird to your 'hospital cage' to give it time to recuperate or you may be able to disguise the injury (if small enough) by using dark coloured sprays such as Gentian Violet.

Gentian violet is a dark purple spray that is anti-bacterial. When sprayed on the bird it very effectively camouflages wounds and helps to prevent further pecking while at the same time helping to clean the wound. Keep the spray away from the eyes of the bird and also from your skin which will be dyed a lovely shade of deep purple. Once when spraying a bird I had the nozzle pointed the wrong way and ended up with a purple speckled face which I don't consider a good 'look'. Only lots of soap and rubbing removed it. However it did not put me off using the product, as it works very well.

F10 barrier ointment is another product that is used to fight bacteria, fungi, viruses, open and contaminated wounds and provides a barrier to help prevent re-infection. Read the instructions carefully, especially with regards to consumption of eggs and meat following the treatment.

As well as treating the injured bird it will be necessary to remove the one that is doing the pecking, and then re-introduce it to the flock after a day or so. You may repeat the process if necessary until the habit is broken. Most importantly go through the list of things that could have stressed the bird (see Pecking Problems Resolved), so you can get to the root of the problem because many pecking incidents arise from the birds being stressed.

CHAPTER 8

HOME CRAFT

This chapter is a collection of small but effective home craft tips that I have accumulated over my time of keeping quail including: how to make Apple Cider Vinegar, how to breed mealworms, how to deepen egg yolk colour and even uses for eggshells.

Amazing Apple Cider Vinegar

Apple Cider Vinegar (ACV) is an excellent overall health tonic for birds and a great addition to their diet. It has many beneficial properties for humans, birds, livestock and pets, but only if you purchase, or make one that contains 'the mother', or 'mother-of-vinegar' which is the beneficial bacteria giving all the health benefits. ACV has a long history of keeping poultry in top condition. It's completely different to other forms of vinegar such as malt or wine vinegar, owing to its medicinal properties.

The Uses of ACV

Administering ACV on a regular basis will offer your birds the following benefits:

Helps to Prevent Intestinal/Faecal Odours

The aroma of the faeces is substantially reduced when ACV is added to their water. It then keeps away flies and makes cleaning out a more pleasant task.

Anti-bacterial and Anticoccidial

ACV contains malic acid, a substance known to have anti-viral, anti-bacterial and anti-fungal properties and so helps to prevent many health problems including preventing coccidiosis.

Anthelmintic (Expels Parasitic Worms)

One great benefit is that it helps to prevent intestinal worms. If your quail are on soil, then they will pick up worms and need standard medicinal treatment. The addition of ACV is an extra back up system, which will maintain your bird's health keeping infestation in check.

Improves Egg Supply

This is a popular claim for ACV. It could be that it overall increases the health and vitality of the birds, which in turn improves their egg production.

Improves Feathering

ACV improves general feather condition. For example even when moulting their feathers will be shiny and luxurious.

Improves Flavouring and Tenderness of Meat Birds

This is a claim for ACV for which I've yet to find evidence, but there is no harm in experimenting in this subjective area. You may find a positive benefit. If you do, please let me know!

Protects from Internal/External Parasites and Flies

A solution of ACV can be sprayed into and around housing as a very effective fly and insect deterrent. Mix a solution of one part ACV to four parts water into a plastic sprayer and treat your housing after cleaning.

Helps to Prevent Soft Shelled Eggs

ACV helps with the absorption of calcium, which in turn helps to form strong eggshells.

Prevents Algae Growth in Troughs and Drinkers

Keep your quails water free from harmful bacteria by using ACV. If your drinking containers are in the sun, in no time at all, they will get slimy and green or black. Keeping them clean is essential to your bird's health, so one of the additional benefits of ACV is that it also prevents algae forming in the containers.

Skin Complaints

A small amount added to bathing water can ease skin complaints at a dose of no more than one part to ten.

Eliminates Mould, Mildew, Dust and Odours

Cleaning with ACV will disinfect any surface and prevent mould, mildew dust and unpleasant odours from building up. Just spray with a solution and wipe clean.

Precautions

Avoid putting ACV in metal drinkers, only use plastic, as the acid in the vinegar will release harmful metallic salts and damage your birds and also rot your metal container.

Avoid giving the birds too much ACV, as too strong a solution could even kill your birds. See below for the correct dosage.

The Correct Dosage of ACV

Use one to two tablespoons of vinegar per gallon of water for maintenance and three tablespoons per gallon for treating sick birds. It's important to note that additional vitamin supplements should not be added in the same drinking water as ACV.

It can be expensive to buy the real un-pasteurised ACV in bulk, but it is very easy to make your own.

How to Make ACV

There are a variety of recipes and methods of making ACV, but for all of them, you need to buy some starter known mysteriously as the 'mother'. 'Mother', which is short for mother-of-vinegar, is the beneficial bacteria that create vinegar. It can look cobwebby, slimy or jelly-like and is the good stuff that gives all the health benefits. To obtain the 'mother', you will need to buy unfiltered, un-pasteurised ACV. It should say on the label that contains 'mother'. This is a small financial outlay, but the benefits are worth it, especially if you continue to make the vinegar on a regular basis.

Ingredients

For the apple content of the vinegar, you have three options: either fresh or frozen apple juice free from additives, chemicals or preservatives. You can make your own juice from autumn apples; summer and green apples don't contain enough sugar to create decent vinegar. Alternatively you may use apple peels and cores left over from food preparation with filtered water and a little sugar. You can also use raw cider (un-pasteurised) for your apple ingredient of the vinegar.

Directions

The quantity you will need to create a new vinegar is about 500 ml liquid ingredient (either the apple juice, or water with apple cores and peels, or raw cider), plus 50 ml of the un-pasteurised unfiltered organic ACV, containing 'mother' to kick-start the process.

Place one chosen apple ingredient (not all three!) into a wide mouthed non-metallic container and cover with muslin or cheesecloth (so air can get in, but bugs and dust cannot). Add some of the ACV that you have purchased *(it must contain mother-of-vinegar)*.

Keep the container away from direct sunlight and maintain the temperature around 15.5°C - 26.5°C (60°F - 80°F) so that the 'mother' will form. Lower temperatures don't always make useable vinegar and higher ones interfere with the formation of the mother-of-vinegar.

Place in a warm room in a dark place away from sunlight (sunlight interferes with the formation of good bacteria). Leave it to stand stirring daily, for three to four weeks with a non-metallic spoon. Stirring daily will add air into the mixture, which increases the rate at which the vinegar will form. Commercially produced vinegars are intensively aerated and can be created in as little as three days.

After about two weeks you should see a gelatinous film on the top of the liquid. This is the beneficial mother-of-vinegar, which is formed when the vinegar bacteria converts the alcohol into vinegar. You should at this point notice an obvious vinegar smell.

You can taste the samples until the desired strength is achieved. The standard acidity is around 5%. You don't need to worry about its acidity unless you are going to use it for preserving foods, in which case you need to purchase a titration kit from a wine or beer making suppliers.

There should be no smell or taste of alcohol present when you are ready to bottle it. Bottle it in plastic bottles with a plastic (not metal) cap. Discard the thick mother-of-vinegar film, or use it to make a new batch. Make sure your containers are spotlessly clean; however it's good to note that ACV does not carry the E. coli bacteria owing to its acidity. It does not need to be refrigerated after bottling, but is best kept in a cool place out of direct sunlight. It has a shelf-life of three to five years.

IMPORTANT

Do not use any metallic spoons, jars, or bottling caps. The acid in the vinegar will corrode the metal and release harmful metallic salts.

ACV showing the cobweb like 'mother-of-vinegar' on the surface

Breeding Mealworms

Quail's favourite snack: fresh or dried mealworms

You can buy dried mealworms from pet shops or online, but you can also breed your own. They work out quite expensive if you buy them, so I've included a guide to breeding your own.

To breed your own mealworms and save lots of money, begin with purchasing some live ones which you can buy in a pet shop or over the internet. Don't worry about buying lots, just buy a few live ones and they will multiply in a short time.

Put them into a high sided container with a lid that has many air holes punched into it. As the beetles don't fly a lid is not necessary, as long as you've got a high sided container, but I have one as an extra precaution and peace of mind.

Next purchase some bran or oats and place them in the freezer overnight or just a couple of hours if you are short for time. This is to kill off any cereal mites that are virtually invisible to the naked eye, but could kill off your colony.

Then, add two-and-a-half centimetres (one inch) of bran or oats. Don't be tempted to add more, as the baby mealworms will get suffocated.

Usually oats work out cheaper, especially if you don't buy them from speciality pet shops. If you buy a supermarkets own brand of oats (use any type) then it will work out remarkably cheap. As a precaution, do not add the feeding material from the worm suppliers as this may contain cereal mites.

Put the mealworms in the tank along with a very small amount of crushed and torn lettuce, green-leafed vegetables, or carrot slices. Any vegetables will do, but from experience I only use leafy vegetables torn up small as anything else goes mouldy and the smallest worms get trapped in it. This is important as they need it as a water source. As the oats or bran is consumed it will be replaced by a darker brown dusty material, so it will need topping up regularly. Do not add too much, as adding moisture can contribute to excellent breeding conditions for mites and mould.

Mites are a common challenge when breeding mealworms. If you get something you think maybe a mite infestation you can identify them by the fact that they move upwards towards the light. They move up the side of the container towards the air holes, whereas baby meal worms of a similar size will not. Also their numbers multiply very quickly. An infestation cannot be remedied and will kill your colony. Consequently the only thing you can do is to kill off the colony yourself. You can do this by freezing them. Before you do so you can remove a number of beetles and set them in a new bucket with fresh food, then move again in one week to ensure they are free from mites.

The best temperature to keep your colony is approximately 21°C (70°F) which usually means indoors, or in a shed in the warmer months. You can speed up production by increasing temperature to 30°C (86°F).

Don't keep them in your pantry or other food storage areas (not that you would!) in the event of any escapes or a cereal mites infesting your food. You should see them go through a complete life-cycle in several months. The lower the temperature in which you keep them, the longer it takes. The worms go hard and whitish then they become a chrysalis, later emerging as beetles. These mate, burrow into the oats/bran and lay eggs. They then die and their bodies are consumed by the next generation of hatching mealworms which are too small to see when first hatched.

To prevent mould growing at the bottom, every couple of months set up a new container and as the worms turn into chrysalis move them to the new container.

Please note that if you find black worms or chrysalis, they are dead. Newly emerged beetles are whitish & turn black or brown as they dry and harden.

Before long you will have lots of mealworms for your quail and wildlife and have saved yourself lots of money in the process.

How to Deepen Egg Yolk Colour

Egg yolk colour can be influenced by what the bird eats. Natural yellow pigments named Xanthophylls influence what shade of yellow the yolks become. For example, feeds based on wheat or barley produce light yellow yolks, those fed on yellow corn or alfalfa meal produce medium yellow yolks. And to get deep yellow yolks pot marigold petals or other naturally occurring orange plant material may be mixed in with their feed. Be careful to choose Pot Marigolds, i.e. *Calendula Officinalis*, not *Tagetes* which are considered by some to be harmful to quail.

Using Eggshells

Nothing need go to waste as you can even make use of your eggshells. A couple of uses are as a slug deterrent and as a calcium supplement.

Slug Deterrent

Eggshells make an excellent slug deterrent as slugs dislike sliding over the rough, spiky and dry surface of the shells. They also blend in nicely by assuming the look of pebbles. To make your slug deterrent simply collect your shells, then microwave them until they are dried out, then crush them until they are quite fine. Sprinkle them around your plants, for an excellent slug deterrent.

Calcium Supplement

Treat the eggshells as above and grind them up finely. You can then add them in a separate container for your quail to eat, to provide extra calcium. Do ensure that you cook them thoroughly. Microwaving is both fast and easy as providing your birds with raw eggshells may encourage them to eat their own eggs, but when cooked and ground up they just see them as a tasty food and something other than their own eggs.

Crushed baked eggshells

CHAPTER 9

MAKING MORE QUAIL

You may come to the point where you would like more quail because your birds have aged, reduced laying, or maybe you would like to produce even more eggs. Now you are going down the slippery slope to quail addiction!

Breeding Basics

To make more quail, you are going to need the following:

- Some fertilized eggs from a reliable source.
- An incubator.
- A brooder with a heat source and thermometer.
- A nursery pen.
- An outside permanent home.

Fertilized eggs - These may be either purchased, or produced from your own stock. Ensure that the birds are unrelated and picked from your best stock, without physical defects, or bad habits such as pecking, or bullying.

Incubator - Needed to hatch the eggs as Japanese quail don't become broody very often. The trait has been diminished over the years and bred-out, while selecting birds for other traits.

A brooder with a heat source and thermometer - This is the place where you place your chicks when they come out of the incubator. It emulates the warmth of the mother hen and they will live here for a number of weeks.

Nursery pen - This is the next level of accommodation before your chicks move outdoors. It is where you acclimatise your chicks to outdoor conditions.

An outside permanent home - This is the final adult accommodation that the fully grown chicks move into when sufficiently acclimatised to the outdoor temperature.

Newly hatched chicks catching taking their first look at the world

An unusual quail: a broody hen with eggs

Obtaining Fertile Eggs

To start creating even more quail you are going to need some birds from which to breed, or purchase some fertilized eggs. Your breeding stock should be your best birds, without defects and the males and females should not be related to each other.

In order to produce a good percentage of fertilized eggs, the usual recommended breeding ratio is to keep your birds in pairs, trios, or one male to five or six females. You can even buy fertilized eggs online and have them posted to you, although the journeys by mail can at times take its toll, so you may wish to order more than you need. Ensure that eggs are collected, or delivered promptly after being laid as fertility diminishes at the astonishing rate of two to three percent per day. Eggs older than a week are much less viable. Check them for hairline cracks and shells with lumps and bumps or other deformities, then discard any that are not perfect. Allow them to settle for a day in a cool area at 15°C (59°F) before setting them to incubate. If you need to keep the eggs longer reduce this to 12°C (54°F), but don't store them in the fridge. Keep them with the pointed end facing downwards, which puts the important air sac at the top blunt end of the egg.

Handling Fertilized Eggs

First unpack the eggs with your hands in protective gloves. This is not for your benefit, but for the benefit of the egg: to prevent transfer of bacteria. Also bring your eggs up to room temperature before setting them, to avoid unnecessarily stressing the fertile eggs.

Washing Eggs

Bacteria from dirty eggs can cause a poor hatch rate, consequently some people wash their eggs in diluted hatchery sanitizer before setting them; others don't wash them at all. If you choose not to wash them and there is dirt present you may gently rub off any visible grime. A nailbrush is ideal for this task. If you do decide to wash your eggs, ensure the water is hand warm as cold water draws bacteria into the egg. Drain and dry the eggs before putting them in the incubator.

Incubators

There is a vast array of incubators on the market. Useful features to look out for are:

- Automatic egg turning.
- Digital temperature control.
- Full humidity control.

Fully automatic incubators are now available at a reasonable price if you are prepared to shop around. Fully automatic means that it automates the time consuming egg turning, temperature control and humidity control. Manual ones involve much more work and risk to the eggs, as you have to turn the eggs 180 degrees a minimum of three times per day, but preferably five times for a total of 14 days, after which all turning stops. It's all too easy to damage the eggs when doing so. If you are just starting out and finances allow, I would advise purchasing an automatic incubator. However, having said that, manual ones still produce chicks very adequately if you are prepared to put in the extra work.

Filling the incubator with the eggs blunt side up

Preparing the Incubator

Prepare your incubator by cleaning with hatchery disinfectant. Next test your incubator by leaving it to run for up to 24 hours. Set the temperature to 37.5°C (99.5°F), humidity at 45%. After a few hours, check the temperature and humidity are working correctly. It's tempting to skip this stage, but it allows you to settle the temperature and humidity, and gain confidence in understanding the controls and instruction booklet. Once the eggs are placed inside there is no opportunity to play around with these without causing damage to the embryo.

The Process of Incubation

Fill your incubator with as many eggs as you wish, placing the eggs blunt side up, which puts the essential air sac at the top. It does not matter if you only partially fill the incubator. Set incubation for 18 days. I am presuming that you will have an automatic turning device in your incubator, but if not mark your eggs with a cross using a soft pencil to indicate which way up they are.

The incubator should be placed in a room with a stable temperature away from draughts, but not outside in a shed or garage as generally the humidity is too high. Consider also that they need to remain stationary for the entire period of incubation. The incubator should also be in an area where it's *not* exposed to vibration. For example avoid the top of the washing machine.

Humidity

Humidity plays a surprisingly important role in the success of incubation. Relative humidity needs to be 45% inside the incubator increasing to 75% when they start pipping (around day 15). During the process of incubation (up to day 15) moisture needs to be evaporated from the egg which will increase the size of the air sac into which the chick will peck, and take its first essential breaths of air into its lungs. After doing this, it takes a long rest to gather its strength for the big push. When the air starts to run out, the chick will begin to peck through the shell.

Quail Egg Air Sac

7th Day

14th Day

18th Day

Increase in air sac size during incubation (approximate)

If the air sac is not large enough, two things may happen: firstly, after resting the chick will not have enough oxygen in the air sac to complete pecking through the shell. Secondly the chick will be unable to rest sufficiently as it has to stand in a difficult position i.e. very upright to peck through the highly situated air sac, using all its reserve of energy. It then becomes exhausted. With both scenarios the outcome is the death of a fully formed chick. This is known as 'dead-in-shell', when doing a post-mortem. Therefore during incubation you may candle your eggs around day six or seven to ensure there is enough air in them. If the air sac is too small, reduce the humidity level to increase its size.

Candling

Candling is basically the process of shining a bright light through an egg when in a pitch-black room, to see if there has been any embryo development and to check the size of the air sac. Another purpose of candling done before you commence incubation is to look for hairline cracks in eggs. Such cracks will prevent embryo development and so you can discard any damaged eggs.

You will need a tool called a candler that is suitable for quail's eggs, as those produced for chicken's eggs may cause confusion owing to the spotted egg markings which can be mistaken for embryo development. With a suitable candler the embryos will show as a shadow and blood vessels.

During incubation candling is usually done on day six when you can check the all important air sac. At this point you may decrease the humidity if the air sac is too small.

Pipping

Around day 15, the quail will begin to peck at the shell to hatch. It's at this stage you should stop turning the eggs. The humidity should be increased to 75% making it easier for the chick to hatch. Increased humidity at this key point keeps their shell membrane soft and prevents it from drying out, making an easier exit from the restrictive shell environment. The temperature should be reduced to 37°C (98.6°F) to compensate for the chick's own body heat production.

'Are you my mum?' A chick's first view of the outside world

Hatching

Chicks typically hatch around day 17 or 18 but they don't all hatch on time. Some are as early as day 14 or as late as day 19. When ready to hatch, the chicks will peck through the membrane of the air sac at the blunt end the egg. They then cut a line of chips out of the shell with their egg tooth. This is known as 'zipping'. Ideally they should cut a neat circular shape around the air sac to make a lid that they then push off. They come out of the shell damp and need time to dry and become fluffy.

A Chick's First Hours

The chicks don't need food as soon as they hatch as the yolk is still in their abdomen, so they can wait 12-24 hours until you put them in the brooder. This allows you to keep the incubator closed to avoid decreasing the humidity levels and killing-off the other emerging chicks by drying out their shell membranes.

Moving your Chicks to the Brooder

Once most eggs have hatched and the chicks have become fluffy, you can move them to the brooder. In order that your chicks don't get a chill, ensure your brooder is warmed up. This can take many hours and is best done in advance of hatching. Chicks are unable to regulate their own body temperature, so it's essential to get this right.

When you're ready to move them, be aware that opening the incubator lid will reduce the humidity which is needed to ease the remaining chicks out of their eggs. This can effectively 'shrink wrap' the shell membrane and kill any remaining chicks that were attempting to hatch out at the time the incubator was opened. Consequently you need to time it as best as you can to avoid fatalities. *Try to avoid removing the chicks when other chicks are pipping (breaking a hole in the shell).*

When you have moved them, once again very quickly check for any eggs that were pipping at the time of the move. These may need to be helped out. To do so, gently peel back a small part of the shell around the hole they have made, until they can escape. If there is a yolk sac still attached to the belly button, leave it attached to both the bird and the shell if possible. If there is any blood when you pierce the shell, leave it alone, as it means that it is too early to emerge. The success rate of assisting in such a way can be hit and miss. Late hatching chicks may be naturally weaker and smaller and they may not survive even if you help them out.

Quail Incubation Check List

Day	Instructions
1	Prepare the incubator by setting the temperature to 37.5°C (99.5°F) and the humidity to 45%. Allow the incubator to warm up for at least two hours. Candle eggs looking for any cracks. Discard cracked eggs. Mark eggs with a cross if using a manual incubator. Place the eggs blunt side up. Note the date and time. Set the incubator for 18 days. Start the incubation period.
2	
3	
4	
5	
6	Candle eggs checking air sac size and embryo development.
7	
8	
9	
10	
11	
12	Check for egg movements, usually a rocking motion.
13	
14	
15	Stop turning the eggs and put them on the hatching mat. Reduce the temperature to 37°C (98.6°F). Raise the humidity to 75% by adding more water to the incubator using the automatic function or adding damp sponges or using small pots with a water guard or anti-drowning pebbles.
16	Check for pipping, usually a tiny raised triangular chip in the eggshell. Listen for cheeping noises.
17	Check for pipping and hatching.
18	Most chicks should have hatched. Transfer to the brooder after 12-24 hours when fluffy and dry.
19	Look out for late pippers and hatchers.
20	Transfer late hatchers to the brooder, and switch off the incubator. Clean the incubator and post-mortem unhatched eggs.

Day old 'Golden' chicks

Brooders

Brooders are your chicks' second home after leaving the incubator. It should be cosy and dry at all times and kept in a warm (not hot) room. Creating your own brooder is quite easy.

An Easy D.I.Y. Brooder

To create your own brooder, you can customise an extra-large plastic storage box, by drilling two rows of large air holes along the top of the box. To do this without cracking the plastic, gently heat the surface of the box before drilling. A hairdryer will warm up the plastic easily and is available in most households, or you may lean the box against a radiator to warm gently taking care not to get it too hot.

An easy to make brooder with: A. Electric Hen B. Non-slip matting C. Paper towels D. Shallow water in a jar lid with small stones to prevent drowning E. Jam jar lid feeder F. Large plastic storage box

Water with small stones to prevent drowning and jam jar lid feeder

The most essential item within your brooder is a heat source. A good system is what is known as an 'electric hen', which is a rectangular heated block that can be raised up on legs. It should be placed so as to give the chicks enough room to walk upright without it touching their heads. Or you can also use an electric light bulb (breeders usually choose red or blue in colour). Chicks gather under the electric hen or bulb to keep warm. They will need such a heat source for around three to four weeks.

Brooder Temperature

The temperature you keep you chicks within the brooder is of paramount importance and you will need to closely monitor it from when the chicks hatch to adulthood. The temperature will need to be gradually reduced to enable the chicks to harden-off ready for outside life. It should be done very gradually over a number of weeks following the temperature guide below, as chicks will die rapidly if it's done too quickly.

These brooder temperatures are approximate and for guidance only. A digital thermometer such as that purchased for poultry or reptile keeping will allow you to monitor the temperature.

Date/Time	Brooder Temperature
First 24 hours	37°C (98.6°F)
Week 1	35°C (95°F)
Week 2	30°C (86°F)
Week 3	25°C (77°F)
Week 4	20°C (68°F)
Week 5	Adjust temperature to adult pen conditions
Week 6	Transfer to adult pen

Observe your chicks' behaviour and adjust the temperature accordingly. Quail chicks cannot adjust their own body temperature by sweating, but regulate their own temperatures by moving around the brooder in and out of the warm zone. Therefore ensure they have a warm area and also a cooler area in the brooder. Observe their behaviour in addition to the number shown on the thermometer; to gauge the well-being of the chicks. If they are gathering all together under the heat source and chirping loudly, then they are too cool. Alternatively if they move to the edges of the brooder and are panting or distraught they are too hot. The ideal situation is for the chicks to be relaxed and moving around the brooder going about their business of eating and drinking, then relaxing again under the heat source.

Ventilation

The chicks should be placed in a well ventilated room that offers a good supply of fresh air. They need the oxygen to thrive however it's not good for them to be in a draughty place that will cause fluctuations of temperature. If they are in a plastic brooder condensation and the build-up of ammonia from faeces can become a problem particularly as the chicks grow. To remedy this you may need a larger brooder and increase the frequency of cleaning. Try to balance warmth with ventilation.

Cold chicks group together for warmth

The Top Ten Features of a Brooder

1. A heat source such as an 'Electric Hen' or light bulb. Please note L.E.D.s are unsuitable as they don't get warm enough.
2. A thermometer to gauge the temperature inside the brooder.
3. A blanket for insulation to increase the temperature of the brooder, when required.
4. A lid for the box to regulate temperature and condensation.
5. Non-slip matting to prevent splaying their legs.
6. Absorbent material to place under the non-slip matting.
7. Bedding material: paper towels for the first few days moving onto absorbent bedding such as dust-free sawdust later on.
8. Water containing marbles/pebbles to prevent drowning.
9. Finely ground quail food.
10. A mesh lid cover/cheese cloth cover to prevent escapees.

Growth

Their rapid rate of growth is truly astounding. Clear changes can be observed on a daily basis with feathers growing within days of their hatch. It is incredible to see them grow from the size of a tiny chick to an adult bird in only weeks.

Only 42 days difference in age: adult female (left) and a day old chick

Bedding Material in the Early Days

Upon hatching you can use simple bedding material to absorb faeces such as thick paper towels. Sprinkle ground up chick crumb leading towards the feeding station, so they can work out where to feed. After about three days or when they all understand where to go, you may put in the non-slip flooring and begin to add an absorbent material on top of any droppings. Virtually any absorbent material is suitable e.g. dust-free wood shavings. Straw is not absorbent and is therefore unsuitable. The absorbent material can be added gradually to cover droppings, but not totally covering the floor at the start in order that the chicks may find their way to the feeders. Once you are confident that they are all feeding correctly, you can increase their bedding material. Be aware that toeballing can occur and their feet should be checked daily and treated appropriately. If you notice toeballing, increase your cleaning routine and use more bedding material.

Cleanliness

Coccidiosis may strike if you don't keep your chick's bedding clean. Bedding should be changed every day, or every other day at the very least. Their tiny lungs may also be damaged with ammonia from their faeces if they are not cleaned out sufficiently. Quite simply, the key to good health for all poultry is cleanliness.

Nutritional Requirements of Quail Chicks

The nutritional requirements of your chicks differ from those of adult quail. With 4% more protein required, 0.05% more methionine, less calcium, and very slightly less phosphorus.

Age and Type	Protein	Methionine	Calcium	Phosphorus
0-6 weeks	24%	0.50%	0.85%	0.6%
Layer from 6 weeks	20%	0.45%	2.75%	0.65%

Appropriate Food Size

Tiny chicks will be unable to eat full size food pellets, so in the first week you need grind adult pellets to a fine crumb. I use a kitchen blender for this process. From week two begin to increase the crumb size until at week three onwards you can begin to offer them full size crumb.

At week three you can begin to offer your chicks the occasional whole mealworm as a treat, but don't overload their tiny digestive systems by giving them too many.

From week four gradually introduce them to some greens and provide them with small sized grit to properly digest their food.

Nursery Pens

At around three to four weeks old once your chicks have grown feathers and start to get larger, you can move them into a nursery pen to gradually harden them off. Something suitable would be a single level rabbit or guinea pig cage made of wire with a removable plastic base for cleaning out. The size will be dependant upon how many chicks you've hatched. Where you put the nursery pen will depend upon the prevailing temperature and time of year. They could remain indoors, or move into a garage, outbuilding or shed, but it is essential that they are kept at the appropriate temperature. They are likely to benefit from a heat lamp or electric hen for a while longer, until they acclimatise. Do ensure you keep your chicks out of any draughts.

Ready for the nursery pen: a three week old female chick

Moving Chicks to Their Outside Homes

At around five or six weeks old the chicks may be moved to their adult accommodation and within days the female birds will begin laying lots of lovely eggs. Once again a sudden change of temperature can adversely affect their health. They will need an artificial heat source for at least the first three weeks or more if the local conditions are very cool.

Making a Plan for Your Male Chicks

You are likely to hatch more male birds than you need for breeding purposes and of course they don't lay eggs, so you will need to decide what is to be done with them. Choices could be to cull them to eat, sell them, or swap them with quail breeding friends to introduce new blood lines. If you would like to keep them please, consider that it's not good to keep all males together once mature as they will fight even to the death. Once they reach maturity bright lighting also increases their rage with one another; so low lighting (not darkness) may help you keep them together longer as a group.

Sexing Quail

Quail may be simply sexed either:

- Comparing their vents.
- Comparing their behaviour.
- Looking at their plumage colours and pattern.

Plumage Differences in Males and Females

From three weeks old the Japanese male quail develop a reddish brown tint on their chest. The females develop spots on their chests and have a paler background plumage. Before three weeks old it's almost impossible to distinguish male from female. And in some strains such as the Texas A&M the plumage does not differ when they reach maturity and you have to rely on the visual differences in the vent area.

The male Japanese quail (left) has a rusty chest, neck and face and the female Japanese quail (right) has spots on her chest

Adult Golden Male (left) showing a deeper rust colour on the chest and face. The female (right) is lighter in colour and has spots on her chest

Vent Differences in Males and Females

Female: the vent (cloaca) sometimes has a purplish tinge

Male: reddened domed vent

Signs of Sexual Maturity in Males and Females

At maturity the male will begin to crow. The sound is *very* much gentler than that of a farmyard cockerel and one that should not disturb you or your neighbours; this of course is ideal for urban farmers.

Once they become adults at around six weeks the male will be interested in mating and will chase any available females. He will also begin to fight any other males that he sees as a threat.

Another sign of maturity in males is leaving foam balls on the cage floor. Foam balls look a little like blobs of shaving cream and are produced by a gland that sits just above the cloaca. This gland secretes large amounts of a thick liquid that contains enzymes and proteins, but it is not sperm (a common misunderstanding). When the male wants to mate he starts to activate a sphincter muscle in the gland which then whips up a foam ball from the liquid. When he mates he puts the foam into her following his semen. The full purpose of the foam ball is as yet unclear. It could provide a good environment for the sperm allowing them to live longer and increase the likelihood of fertilization, or it may help the sperm towards its destination.

When a female matures she will begin to lay eggs at around 42 days old and she does not fight aggressively like a male. A group of females do however establish a pecking order, but under normal circumstances it is less serious than the fight-to-the-death attitude of the males.

Foam ball produced by male quail marking the covey floor

Foam ball produced by a male quail (right) next to an egg

Feather Colour and Genetic Dominance

If you have gone along the path of breeding some quail you may be interested in developing your own special genetic line. If so it's worth considering the following colour and feathering traits and their dominant and recessive properties.

Colouring: Dominant and Recessive Genes of Quail

Dominant	Recessive	Incompletely dominant
Black at hatch (Bh) Dominant over all colours, but should not be bred together owing to eggs very low hatch rate.	Cinnamon (cin) Recessive gene giving ginger colour feathers.	Extended Brown (E) This colour is incompletely dominant over the normal type. The birds are brown all over.
Yellow (Y) Produces 'Golden' birds. Fatalities of 25% if bred together, but if you breed it with a brown, you will get mostly goldens and some brown in the proportion 2:1.	White Breasted (wb) Produces white-feathered breast, neck and face in males and females.	Silver (B) If bred together they show slow growth and slow sexual maturity.
	Redhead (e^{th}) This gene produces white birds, with irregular black and rust feathering.	
	Imperfect albinism (al) Two imperfect albinos produce totally albino chicks with pink eyes and white feathers.	

Lethal Combinations

Sometimes interbreeding produces unintended consequences and the following the results are lethal: *Black X Black and Golden X Golden.* Consequently it's best to avoid breeding these colours together. Some may survive but you can expect high fatalities and health complications.

Feather Mutations

There are also mutations of the feather that may occur from time to time and should be considered when you embark upon breeding your own quail.

Porcupine (pc)

Recessive - Giving abnormal curled feathers on the back.

Rough textured (rt)

Recessive - Giving feathers that are rough to the touch.

Ruffle (rf)

Recessive - Giving soft barbs to some of the feathers.

Short barb (sb)

Recessive - Giving the appearance of short and broken back feathers.

Defective feathering (Df)

Dominant - Giving short, meagre feathering.

With the knowledge of colour and feather genetics it's possible to develop strains that move away from the traditional type.

CHAPTER 10

KEEPING QUAIL FOR MEAT

Quail meat is a delicacy often associated with the high life and rightly so, as it's a delicious food. If you have never eaten quail, imagine the tastiest chicken you have ever eaten, then intensify the flavour.

The Benefits of Producing Your Own Meat

If you raise your own quail for meat, you will have no doubt as to their origins, management, feed or life quality, as you have been responsible for it all along with the non-existent 'food miles'. And like keeping quail for eggs, it is very easy to feed them with organic produce without costing a fortune. In no time at all, you can be producing your own organic meat for very little cost.

Selecting Birds to Eat

Many birds that are eaten by small-scale breeders are the mature males left over from hatching that are obviously unable to lay eggs and not picked as prime specimens for further breeding programmes. These birds are usually kept all together and separate from the rest of the flock and eventually culled and used as table birds. Unwanted female birds are equally desirable to eat.

As males have a tendency to fight when they mature at around 42 days, a group of male only birds will need to be managed properly as the fighting is vicious, bloody and fatal.

You will need to judge the time to cull ideally before they begin to fight, but when they are at a good body weight.

'When They Start to Crow, It's Time to Go!'

They will reach maturity at six to eight weeks old. Some birds mature quicker than others and will begin crowing and racing around fighting while others are still immature. The problem with them racing around means they will actually lose weight and not make a good weight to eat, therefore you need to take measures to keep them calm. You can also make their tendency to fight much worse by keeping them in a strongly lit environment. Slightly subdued light (not total gloom) keeps the males calmer when approaching maturity and allows them to put on weight instead of racing around trying to kill each other. However, as a general rule, *'When they start to crow, it's time to go!'*

First cull some of the mature male birds, but not the less mature ones who make take a few weeks more to reach maturity and a good body weight for eating.

Breeds for the Table

Certain birds are bred specifically for meat production, such as the Texas A&M and Jumbo strains of *Coturnix*. The goal of selectively breeding these birds has been to produce a larger carcass with more meat on the breast and legs and less emphasis upon egg production. In home meat production, if you have these strains of bird, there is little difference in rearing them to their egg laying sisters. You may even get an egg or two as a bonus. See the 'Quail for Eggs' chapter which outlines the different strains and their uses.

How to Achieve a Constant Supply of Meat

How many quail do you need to get a constant supply of meat? Are you ready for more quail mathematics? Before you go ahead please note this calculation assumes a good fertility rate and no fatalities once they are hatched. As fertility is variable and fatalities do occur, the calculation is purely academic. However it should point you in the right direction to begin producing roughly correct quantities for your needs.

The Eggs to Meat Calculation

If you hatch a hundred quail eggs, and get a good hatch rate of say 75% which is 75 birds, your next move would be to then keep 25 of the females and four males for further breeding and laying of eggs. This would give you 46 birds for the freezer, which is enough for 23 individual meals (2 quail per person). In this calculation 46% of the quail hatch is produced for meat.

The remaining birds could be used for egg and meat production. You can then collect around 25 eggs per day (one egg per female hen) which amounts to 175 eggs per week to continue hatching and repeating the process once your freezer stock gets low. Remember it takes a minimum of around 42 days for them to be ready for culling.

So given this calculation you can make an educated guess at the numbers of eggs you will need to hatch to produce all your home reared meat requirements.

Selection for Breeding

For continued breeding you need to select your best birds, without defects and ideally choose those with the highest weights at 42 days in order to continue breeding the heavier weight for the table.

Light Restriction

Some breeders restrict light in an attempt to slow down the rate at which the birds reach sexual maturity, thinking that this will increase carcass weight. However apart from being inhumane, the verdict is out as to whether it actually works or not.

What Age to Cull?

As a general rule quail can be eaten from around six weeks old. They are at their best at around eight to ten weeks and after that they grow gradually tougher. At around only 8 months old they are on the brink of being inedible. Older birds may be used for creating a soup stock or feeding pets if you don't want to waste them.

You've heard the phrase 'She is a tough old bird' and this rings true. The phrase took on a new significance when I decided not to waste an 18 month old hen. I slow cooked it for several hours, thinking the meat would be falling off the bone in that time. However, it was as succulent as chewing Blu Tack and was fed to my dog. In our world, where chickens are slaughtered at the right age to be soft and succulent, it can come as a surprise when you first try to sink your teeth into the un-yielding flesh of an old hen, even when well prepared and most carefully cooked.

Selling Quail Meat

Once you move beyond the realms of providing meat for your own table there are strict rules and regulations regarding the sale of quail meat. DEFRA will provide you with the relevant information and should be consulted if you decide to go down this route. It is also advisable to contact your Local Authority Environmental Health Department, to discuss whether or not premises need to be inspected and how to dispose of processing by-products.

CHAPTER 11

CULLING

Culling is not a very pleasant subject, however, it's something all poultry keepers have to deal with and it helps to be informed of the most painless and up-to-date methods. It is of course a very tough decision to make and we may procrastinate, to the detriment of the quail in question who may suffer far too long, while we put off the inevitable. When the time comes, we need to cull our birds in the quickest, easiest and kindest way possible.

When you get to the point of having to dispatch one of your birds, there are a few methods to consider. You can of course go to the vets to have your bird put to sleep. However this is expensive. If you can afford it, then it's probably easiest on your sensibilities. However, most of us have to resort to doing it ourselves. As the person who has reared your birds, there is no one better than you, the person who has raised and cared for them.

Don't waste time beating yourself up about having to do this; if you are dispatching a bird because it's too sick to carry on, then you are ending its misery. And from an ethical stand point you are neither cruel, nor 'sick', for culling your quail in a compassionate manner.

Principles for a Painless Cull

Many keepers of small poultry flocks will not be able to afford the present cost of stunning equipment which is currently advocated as part of the most humane method. Until these items are reduced in price, it's likely you will have to rely on other methods. To help you decide, it's essential to note:

- The methods that involve crushing the bird's vertebrae are not as humane as those that stretch the neck as a whole; as stretching also over extends and damages the blood vessels.
- These blood vessels keep the brain oxygenated and therefore keep the bird conscious, so stretching them will cut off the oxygen supply to their brain, and the bird becomes unconscious.
- Merely using a method that crushes the vertebrae will keep the brain alive for longer and is therefore less humane.

The Best Culling Methods Without a Stunner

Method One

This method is commonly used to butcher quail: firmly hold the body of the quail with one hand and swiftly pull hard to remove the actual head of the bird with the other hand. Although it may appear to stem from the Ozzy Ozborne School of Farming, it works by stretching the neck damaging the blood supply to the brain and avoids crushing the vertebrae. On first hearing this, your reaction to the thought of removing the head of one of your flock may seem barbaric. By understanding the fact that stretching the neck damages blood supply to the brain and that first the bird loses consciousness, helps the understanding that this is much kinder than crushing its neck and keeping its brain conscious. To complete the cull swiftly, causing the bird minimal pain, the head must be pulled off in one swift, strong, jerking action, with the strength of something like trying to fight with a door knob. Then the body should be placed downwards, so that the blood can be drained off. It can help to see someone else do this before you attempt it yourself, or look online for videos showing culling using this method.

Method Two

For this method you hold the quail in one hand and use very sharp poultry sheers to remove the head. You can hold the bird upside down and allow the blood to drain away. By comparison to the head removal method, this manner does not stretch the blood vessels of the neck and so the brain of the bird may be alive for some time. So, although method one appears more barbaric, it's actually more humane. *I would therefore recommend method one, in the absence of any stunning equipment.*

The Moments After Culling

Whichever method you may use, be aware that the body will twitch vigorously afterwards for around 15 seconds. This is just muscular spasms and does not mean the bird is in pain. Allow a few minutes before processing the carcass as the heart will beat for a little while, which although the animal is dead, can be very disconcerting if you begin processing too early.

The Location of the Cull

One point than can be overlooked is to do the task far away from your flock of quail. Locate yourself down wind, around 30 feet or more, where they will not see the dispatching, nor will they smell any blood. If they do smell it, it will cause them to panic; which will also result in them going off lay for a while. The question of no eggs for a period of time is of course far less important than the unnecessary suffering of your flock. The smell will travel to them a surprisingly long way, so take your bird to a quiet place far away from your birds to complete the procedure. It's also not a great procedure for your neighbours to see!

There can be quite an amount of blood, for such a small bird and it can be helpful to complete the process using a large plastic bag and kitchen roll beneath the bird to catch avoid any spillage of blood.

Up-to-Date Advice

Get the best advice currently available, as practices change and hopefully improve and electric stunners are reducing in price. The Humane Slaughter Association offers superb advice regarding culling. Its leaflet *Practical Slaughter of Poultry - A guide for the Small Producer* gives up-to-date information on the most humane methods. There is a nominal fee for the booklet, which was quickly sent and well worth the small investment for the excellent advice and format. You can visit their website at www.hsa.org.uk

Tips

- Plan ahead.
- Comply with the law.
- Be quick and confident in the method you choose and do not start a task you cannot complete.
- Check for signs of an effective kill and don't hesitate to repeat the slaughter technique if unsure/necessary.
- Don't cull in the gaze of your neighbours!
- Don't cull in front of your other quail.
- Don't cull where your quail can smell the blood.

CHAPTER 12

PREPARING QUAIL FOR THE TABLE

It's relatively easy to prepare the birds for eating. You can choose to either pluck the bird which is quicker than you may imagine, or skin it which is even faster.

How to Process the Carcass

Once culled it is better to process the dead bird quickly before they get cold. Process the bird in an area that is spotlessly clean, away from all other food stuffs. Whether you are plucking or skinning you will need to remove the head if it is still present using poultry shears, and discard. Then follow the method for either plucking or skinning.

Plucking

To loosen and relax the feathers dip the bird in boiling water for a few seconds only (any longer and you will start to cook it). Submerge until the wing feathers may be easily removed. Then immediately dip the bird in ice cold water and then continue to remove the rest of the feathers. With this method the skin remains on the quail meat.

Skinning the Quail

First cut off the legs at the 'knees'. Then snip off the wings right at the side of the body (there is virtually no useable meat on them). Next put your fingers inside the hole where the wings were and pull off the skin and feathers. It comes off quite easily like taking off a jacket and more or less in one piece. Finally remove any remaining feathers.

Removing the Innards

An easy way to remove the innards is to starting cut from the bird's vent using poultry shears. Cut up either side of the spine from bottom to top. Then hold the remaining neck and pull it out, which should then remove the unwanted innards all in one piece and quite easily. If there are some innards remaining, use a teaspoon to scrape them out. Discard all the unwanted innards and then rinse the bird in cold water and prepare for storage or cooking.

CHAPTER 13

COOKING WITH QUAIL EGGS

Quails' eggs are delicious and easy to replace chickens' eggs in any recipe. You may interchange them, remembering that around five quail eggs equates to one chicken egg. The following pages contain some of my most favourite recipes and tips on handling quail eggs.

How to Crack Quail Eggs

Cracking eggs for use in cooking often causes difficulties in the early days as you get used to their small size and thin shells. There are a few methods to break them easily, for example you may use a sharp knife to crack the shell at the thick end or middle of the egg. Alternatively you may crack the side of the egg gently against the side of a cup.

Using either method you will from time to time break off a splinter of eggshell in your cracked eggs mixture; which you can remove using a spoon if you want the eggs whole; or you can remove the largest pieces of shell then whisk the eggs and sieve the mixture, which is great for items like scrambled eggs, cake mixtures.

How Long Does It Take to Hard Boil a Quail Egg?

Hard boiled eggs can be eaten alone, or used in other recipes such as miniature Scotch eggs, or pickled eggs. Please note that very fresh eggs are harder to shell by hand than eggs that are a few days old as the inner membrane sticks to the shell more tenaciously in very fresh eggs.

- Firstly check for cracked eggs and remove any that you find.
- Soak your eggs for a little while in cold water to clean them.
- Place your eggs in a pan and cover with cold water.
- Bring it to the boil and then set a timer and simmer, for *five minutes*.

If you are going to use the eggs in pickled egg recipes, turn the eggs every minute when simmering, so that they do not set with the yolks lopsided or sticking to the outer membrane.

Take the eggs out of the water and put into cold water for five minutes. This will stop there being a dark ring around the egg yolk.

Once cooled you can start peeling (for easy peeling instructions see below).

Store in the refrigerator until ready for use.

How Long to Boil an Egg to Achieve a Runny Yolk?

Lightly boil for *exactly two minutes thirty seconds*. Don't guess at the timing as the yolk's solidity changes rapidly.

Easy Methods for Shelling Quail Eggs

It's important to note that shelling fresh eggs under a week old is very much harder than shelling older eggs, as their membrane sticks more tenaciously and chunks of the egg are removed at the same time. Once over a week old it becomes much easier to shell them as the membrane loosens its grip.

To assist with the removal of the shell you can plunge freshly boiled eggs into a container of water with a good number of ice cubes and leave it for half an hour. The shells will then come off more easily.

Another method is to soak the egg in a solution of 100% clear 'white' vinegar for approximately an hour until the shell becomes softer. Then make a hole at the blunt end of the egg, where there will be an air bubble. This makes it easier to rip off the rest of the rubbery shell in one or two goes.

To totally dissolve the eggshells with virtually no effort at all, just place the egg in vinegar solution. It takes around 12 hours and the shell will dissolve totally leaving just a rubbery membrane which you simply rub off the egg.

When you soak the eggs in vinegar you will notice that first the coating of spots float to the surface, leaving you with a pale bluish white shell. This is followed by a dissolving of the shell some hours later. You are then left with a soft rubbery membrane around the egg, which you will need to remove by hand. Leaving them in this solution for 12 hours will impart a vinegar taste, which is great for pickled eggs but not a lot else. Put a lid on the pan, to avoid acid vinegar aromas that will permeate your house.

CHAPTER 14

QUAIL EGG RECIPES

Quails' eggs may be used as a replacement for chickens' eggs by exchanging approximately five quail eggs to one chicken's egg. Their small size and softer texture enhances virtually every recipe I've tried. I've listed for you a few of my top favourite recipes that have been tried and tested over a number of years.

Thai Spiced Caramelized Quail Eggs (Kai Look Kui)

This is a delicious starter dish for two people with a sweet and sour flavour.

Ingredients

- 7 quail eggs
- oil for frying
- 3 red chillies (or to taste)
- 2 cloves of garlic (or to taste)
- 1 teaspoon of chopped coriander leaves
- 1 tablespoon of shallots or red onion, finely chopped
- 2 tablespoons of light soy sauce
- 2 tablespoons of Nam Pla (fish sauce)
- 1 tablespoon of vinegar (white wine, or rice wine vinegar)
- 1 tablespoon of sugar (brown sugar will do, but palm sugar is the authentic choice)

Directions

1. Hard boil the eggs as above and peel.
2. Fry the whole boiled quail eggs gently until they change colour slightly and place on the serving plate.
3. Fry the shallots or red onion until brown but not burnt and put on top of the eggs, which can be sliced open or left whole.
4. Slice the chilli and garlic and fry in the oil for a minute to release the flavours.
5. Turn the heat to medium and add the soy sauce, fish sauce, sugar, and vinegar and fry until the sugar caramelizes and the sauce thickens and becomes brown.
6. Pour the sauce over the eggs.
7. Sprinkle over the chopped coriander leaves.
8. Serve with sliced cucumber, tomato, salad and rice.

Quail Eggs with Dukkah

This is a tasty starter of boiled quail eggs with Middle Eastern spices. It would be suitable on its own, or as part of a Meze (a collection of hot and cold dishes, served together at the beginning of a meal in the Middle East, Greece, Turkey and the Balkans).

Ingredients
    18 quail eggs
    sea salt

For the Dukkah
¼ cup sesame seeds
¼ cup roasted whole blanched almonds
2 tablespoons cumin seeds
2 tablespoons coriander seeds

Directions
1. For the Dukkah, roast all the seeds and nuts until golden brown but not burnt. Roast them separately as they take differing times to cook. Leave them to cool.
2. When cool, place both the seeds and nuts into a food processor and blend until they become a rough paste.
3. Season with the sea salt and transfer to a mixing bowl.
4. Boil the quail eggs for five minutes.
5. Peel the eggs.
6. Serve with a small salad and/or other Meze items, or on a wooden board or plate with a bowl of Dukkah in the centre for dipping.

Quail Eggs in Aspic

There is nothing complicated about this recipe, but it makes a great impression and may be prepared in advance. Aspic is basically gelatinous clarified stock, made from meat or chicken, with added herbs and spices.

Serve with smoked salmon and a salad.

Ingredients
- 1 dozen quail eggs
- ¾ oz (20 g) aspic powder
- 2 tablespoons of white wine
- finely sliced red peppers, cucumber skins and watercress (optional).

Directions
1. Boil the quail eggs for five minutes and then shell them.
2. Create your aspic solution and add the white wine and allow it to cool.
3. Pour a very thin layer of aspic into four ramekin dishes. Leave to cool.
4. Optional: to make a flower decoration that sits inside the aspic, cut small flower petal shapes from the red peppers and very thin strips of cucumber skin to create the flower stalks and leaves.
5. Dip the decorative strips of vegetables into aspic and then arrange your flower pattern on top of the set aspic.
6. When the vegetable decorations have set and stuck to the previous layer add the tiniest amount of aspic to hold the flower shapes in place (around 15 ml of aspic) and allow to set again.
7. Place three eggs into each dish, cover with aspic and allow to set.
8. To serve, dip the ramekins in boiling water for a few seconds. Any longer and the aspic will melt. Turn out and garnish with watercress and flower shapes from the remaining peppers.

Courgette Frittata

This is an easy recipe that is very tasty and you can eat it hot or cold. It is also ideal as a breakfast dish.

Ingredients

- 4 tablespoons extra-virgin olive oil
- 1 lb (500 g) of thinly sliced courgette
- 72 quail eggs
- 1 teaspoon salt
- 2 teaspoon dried parsley
- 1 oz (28 g) grated parmesan cheese
- 2 teaspoons dried garlic

Directions

1. In a thick based skillet over a medium-high setting, heat two tablespoons of the olive oil until hot.
2. Add the courgette and sauté, stirring frequently until golden brown and slightly limp, which should take around 6-8 minutes.
3. Remove them from the skillet and allow the courgettes to cool slightly.
4. Shell and then whisk the eggs, then sieve them to remove any shell particles if present.
5. Add the salt, parsley, Parmesan cheese and garlic in a large bowl and stir in the courgette.
6. Heat the remaining olive oil in the skillet over a medium-high setting.
7. Add the courgette-egg mixture and cook without stirring until the bottom and sides begin to set. This should take around four minutes.
8. Transfer the skillet to the oven and bake until the eggs completely set and the top is golden brown, which should take around 15 to 20 minutes.
9. To serve, set a plate on the top of the skillet (the plate's underside should face up) and holding plate to skillet with both hands, gently flip over. Lift off the skillet.
10. Slice into wedges and serve.

Traffic Light Quail Eggs

See a photograph of this dish heading chapter 14. It is an easy and colourful supper dish to make.

Ingredients
- 4 red, yellow and green bell peppers sliced into rings of medium thickness
- 20 quail eggs

Directions
1. Put the bell pepper rings into a frying pan and lightly cook for five minutes.
2. Crack the quail eggs into a jug.
3. Turn down the heat pour the quail eggs into the rings.
4. Cover until set.
5. Serve with a side salad.

Mini Scotch Eggs

Ingredients

- 1 lb (500 g) pork mince
- 2 teaspoons sausage spices
- 2 ½ fl oz (75 ml) Water
- 16 hard boiled quail eggs
- 3 ½ oz (100 g) dried bread crumbs
- 1 peeled and grated apple (optional)

Directions

1. Mix the pork mince and grated apple with the water and mix together with your hands until it resembles sausage meat.
2. Leave overnight if possible for the spices to permeate the meat.
3. Coat the pork mixture around the eggs.
4. Roll the pork and egg balls in the bread crumbs.
5. Spray with an oil spray and bake in the oven for 35 minutes at 200°C (390°F) or deep fry until cooked through.

Pickled Quail Eggs

Excess quail eggs may be pickled to store them, or to sell, or as great Christmas stocking fillers.

Ingredients

    25 boiled and shelled quail eggs
    12 fl oz (350 ml) of clear pickling vinegar
    ¾ teaspoon dill seed
    ¼ teaspoon white pepper
    3 teaspoon salt
    ¼ teaspoon mustard seed
    ½ teaspoon onion juice
    ½ teaspoon minced garlic

Directions

1. Follow the previous instructions for hard boiled quail eggs (see Cooking with Quail Eggs) then remove their shells.
2. Rinse the eggs and peel them.
3. Rinse a second time and put into a sterilised jar.
4. Place the pickling ingredients into a pan and simmer for 15 minutes.
5. Pour the pickling liquid into the jar, completely covering the eggs and seal.
6. To allow the spices to permeate the eggs, allow the sealed jars to sit for at least two weeks before eating.

Hot Chilli Pickled Quail Eggs

Follow the previous recipe, but use the following spices instead.

Ingredients
½ teaspoon dill seed
2 tablespoons of diced onion
3 teaspoons salt
2 teaspoons dried garlic
4 tablespoons of Tabasco sauce (or to your own taste)
2 teaspoons sugar
1 teaspoon whole dried cumin
¼ teaspoon brown mustard seed

Directions
Leave the eggs to absorb the flavours for two weeks or more.

Easy Ice Cream

This recipe is a top family favourite, which unlike many ice creams does not require any beating when going through the freezing stage.
Separating whites from yolks takes time and care, but is well worth it. If you cannot face that delicate task, use larger top quality chicken's eggs. Please note the recipe contains raw eggs.

Ingredients
- 12 quail eggs, separated
- 2 oz (56 g) icing sugar
- ¼ pint (140 ml) double cream (use a good local source of cream for a superb flavour)
- 1 teaspoon vanilla extract (not essence)

Directions
1. Separate the egg yolks from the whites very carefully.
2. Beat the egg whites until starting to peak.
3. Slowly add the icing sugar.
4. Lightly beat the egg yolks and add the vanilla essence.
5. Beat the cream until starting to peak.
6. Add the cream to the yolk mixture. Please note, the more you whisk the eggs and cream, the harder the ice cream will be.
7. Freeze.
8. There is no need to stir or whisk the freezing mixture (unlike many other ice cream making).

Christmas Ripple Ice Cream

My family were still talking about this wonderful ice cream many months after eating it at Christmastime. It was a huge hit and can of course be eaten at any time of year.

Directions

Follow the recipe above for the basic ice cream mixture, but add six tablespoons of fruit mincemeat (or to taste) to the recipe and swirl it through the mixture just before freezing. Homemade mincemeat is best of course, but good quality shop bought will do, if you are in a hurry.

Muffin in a Minute

This is a top favourite snack that can be made in a hurry and is great for children to make. It's very easy to make and makes a great breakfast muffin. It has the added benefit of being gluten free and also low glycemic load, if a sweetener such as xylitol is used instead of sugar. The cornmeal also has a lower glycemic load than ordinary flour.

Because of the small quantities, it's very important to measure the ingredients accurately.

Ingredients
½ oz (18 g) fine ground yellow cornmeal (fine polenta)
2 quail eggs
2 teaspoons natural yoghurt
½ teaspoon natural flavouring such as vanilla, butterscotch, coconut
¼ teaspoon cinnamon
½ teaspoon baking powder (be extra careful at measuring this ingredient)
1½ teaspoons sugar or substitute such as xylitol
¼ oz (10 g) peanut butter

Directions
1. In a small breakfast cup, stir all the ingredients except the peanut butter together until thoroughly mixed.
2. Add the peanut butter to the top of the mixture, but do not stir.
3. Microwave for 60 seconds (1000 watt).

Serve with fresh yoghurt or cream.

CHAPTER 15

QUAIL MEAT RECIPES

I first ate quail many years ago in Turkey, when I was offered one that had been cooked outside upon on a charcoal grill. I was not impressed with its small size, but out of curiosity I decided to try it. Upon eating it, I was very pleasantly surprised and could not imagine why they were not more widely eaten as it was far tastier than our usual chicken meat.

The simple but very tasty Turkish quail recipe is the first one listed in this chapter and can be cooked on a normal grill or barbecue.

Simple Turkish Style Roast Quail

In Turkey the quail are usually cut open and flattened before cooking.

Serves four

Ingredients
   8 quail
   salt and pepper to taste

Directions
1. Open the quail by cutting along the backbone with poultry shears.
2. Remove the spine and ribs.
3. Flatten the meat and season.
4. Place on a moderately hot barbecue four to six inches (10-15 cm) above the coals.
5. Grill on both sides until cooked through.
6. Serve with rice and a mixed salad.

Egyptian Style Roast Quail

Since ancient times quail have been eaten in Egypt. This simple recipe is a traditional, easy and tasty way of preparing them.

Serves four

Ingredients

8 quail
30 ml lemon juice
3 tablespoons of extra-virgin olive oil
1 tablespoon of ground cumin
salt and fresh ground black pepper
250 ml (8 fl oz) chicken stock

Directions

1. Wash the quail and pat dry with a paper towel and season with salt and pepper.
2. Mix together the olive oil, lemon juice, cumin, salt and pepper in a large bowl and mix well.
3. Add the quail to the mixture and stir, ensuring the mixture covers all of the quail.
4. Cover the bowl in cling film to marinade overnight.
5. Pre-heat the oven to 180°C (350°F).
6. Pour the chicken stock into a roasting pan.
7. Add the quail, breast side up.
8. Pour the remaining marinade over the quail.
9. Roast for 25 minutes, until the juices run clear.
10. Remove the quail to warm plates.
11. Place the pan onto the top of the stove and heat the juices over a medium heat for two minutes.
12. Surround the quail with some gravy.

Serve with rice and cooked vegetables.

Honey and Bourbon Marinated Quail

This is a sweet American style dish. Bourbon lends itself to the mild gaminess of quail meat and is very tasty prepared this way. In no time at all your meal will be quickly reduced to a pile of bones.

Ingredients
- 4 quail
- 1 tablespoon bourbon
- 4 tablespoons runny honey
- 2 tablespoons olive oil (not virgin olive oil as it will overwhelm the other flavours)

Directions
1. Separate the quail into joints by cutting up along each side of the back bone with poultry shears from the vent to the neck to make two halves.
2. Place all the ingredients except the quail into a zip lock sandwich bag.
3. Add the quail and marinade for several hours or overnight.
4. Bake on 180°C or 350°F for 25 minutes until the juices run clear.

Almond and Polenta Herb Crusted Quail

A mild and tasty dish with almonds adding extra moisture to the quail meat.

Ingredients

4 jointed quail
2 oz (55 g) whole almonds or hazel nuts
1 oz (30 g) cornmeal or 2 oz (60 g) breadcrumbs
5 quail eggs
1 tablespoon olive oil
½ teaspoon dried garlic
½ teaspoon mixed herbs
½ teaspoon salt
ground pepper

Directions

1. Put the nuts into a food processor and chop until they are in small pieces, but not full ground up (it should resemble an ice cream topping).
2. Mix the nuts with the cornmeal or bread crumbs, dried garlic, mixed herbs, salt and ground pepper.
3. Drizzle the olive oil into the crumb mixture.
4. Whisk the eggs.
5. Dip the quail joints into the whisked eggs and then coat the crumb mixture onto the quail.
6. Bake for 15 to 25 minutes until cooked through and the coating is crisp 180°C (350°F).

Smoked Cheese and Apple Stuffed Roasted Quail

Quail meat is complimented by being smoked. You can smoke them separately or simply add smoked ingredients as in this recipe.

Serves four

Ingredients

8 whole quail
8 slices smoked bacon
4 oz (120 g) smoked cheese
2 apples – grated
1 tablespoon flaked almonds
½ teaspoon dried garlic
½ teaspoon mixed herbs

Directions

1. Mix the cheese, almonds, apple, garlic and herbs.
2. Stuff the cavity of the quail with the stuffing.
3. Wrap each quail in one piece of smoked bacon.
4. Bake for 25 minutes until cooked through on 180°C (350°F) until the juices run clear.
5. Allow the quail to rest for five minutes in a warm place to relax the meat.
6. Serve with mashed potatoes and mixed vegetables.

Quail Tandoori

Indian Restaurant Tandoori Chicken pales in significance when compared to this dish. The strong spices of the Tandoori don't seem to get in the way of the flavour of the quail, only enhance it.

Ingredients

    2-3 lb (1-1.5 kg) whole quail or a mixture of breasts and legs joints.
    1 teaspoon salt
    1 lemon
    1 clove of garlic – crushed
    9 fl oz (250 ml) plain natural yoghurt
    half a green chilli finely chopped or half a teaspoon of chilli powder
    1 small cube of fresh ginger- crushed
    2 teaspoons garam masala
    1 teaspoon paprika
    ground black pepper

Directions

1. Wash and pat dry the quail.
2. Skin the quail.
3. Score the flesh of the quail with small marks once or twice against the grain of the flesh.
4. Sprinkle the quail with salt and squeeze lemon juice into the slits.
5. Mix the remaining ingredients together.
6. Marinade for 4-6 hours or overnight if possible.
7. Arrange the quail on a baking tray.
8. Bake for 20-25 minutes 180°C (350°F) turning once during cooking.
9. Ensure that the juices run clear and the meat is cooked through.
10. Serve with basmati rice and side salad.

APPENDICES

Appendix A: OLD WORLD QUAIL

Order: GALLIFORMES Family: Phasianidae (Pheasants and Allies)		
IOC English Name	**Scientific Name**	**Range**
Common Quail	*Coturnix coturnix* (Linnaeus, 1758)	AF, EU : Europe and c Asia
	C. c. coturnix (Linnaeus, 1758)	Europe and nw Africa to Mongolia and n India
	C. c. conturbans (Hartert, 1917)	Azores
	C. c. inopinata (Hartert, 1917)	Cape Verde Is.
	C. c. africana (Temminck & Schlegel, 1849)	sub-Saharan Africa, Mauritius, Comoros Is., Madagascar
	C. c. erlangeri (Zedlitz, 1912)	e and ne Africa
Japanese Quail	*Coturnix japonica* (Temminck & Schlegel, 1849)	EU : Mongolia and e Siberia, Japan and Korea
Rain Quail	*Coturnix coromandelica* (Gmelin, 1789)	OR : Pakistan to Sri Lanka and Burma
Harlequin Quail	*Coturnix delegorguei* (Delegorgue, 1847)	AF : widespread
	C. d. delegorguei (Delegorgue, 1847)	Africa south of the Sahara, Madagascar
	C. d. histrionica (Hartlaub, 1849)	São Tomé
	C. d. arabica (Bannerman, 1929)	s Arabia
Stubble Quail	*Coturnix pectoralis* (Gould, 1837)	AU : se, sw Australia
New Zealand Quail	*Coturnix novaezelandiae* †(Quoy & Gaimard, 1832)	AU : New Zealand
Brown Quail	*Coturnix ypsilophora* (Bosc, 1792)	AU : widespread
	C. y. raaltenii (Müller, S, 1842)	Flores, Timor and adjacent islands (Lesser Sundas)
	C. y. pallidior (Hartert, 1897)	Sumba and Sawu (Lesser Sundas)
	C. y. saturatior (Hartert, 1930)	n New Guinea lowlands
	C. y. dogwa (Mayr & Rand, 1935)	s New Guinea lowlands
	C. y. plumbea (Salvadori, 1894)	ne New Guinea lowlands
	C. y. monticola (Mayr & Rand, 1935)	alpine New Guinea
	C. y. mafulu (Mayr & Rand, 1935)	montane New Guinea
	C. y. australis (Latham, 1801)	Australia
	C. y. ypsilophora (Bosc, 1792)	Tasmania

King Quail	Excalfactoria chinensis (Linnaeus, 1766)	OR, AU : widespread
	E. c. chinensis (Linnaeus, 1766)	India to Sri Lanka, Malay Pen., Indochina, se China and Taiwan
	E. c. trinkutensis (Richmond, 1902)	Nicobar Is.
	E. c. palmeri (Riley, 1919)	Sumatra, Java
	E. c. lineata (Scopoli, 1786)	Philippines, Borneo, Lesser Sundas, Sulawesi, Moluccas
	E. c. novaeguineae (Rand, 1941)	montane New Guinea
	E. c. papuensis (Mayr & Rand, 1936)	se New Guinea
	E. c. lepida (Hartlaub, 1879)	Bismarck Arch.
	E. c. australis (Gould, 1865)	e Australia
	E. c. colletti,(Mathews, 1912)	n Australia
Blue Quail	Excalfactoria adansonii (Verreaux, J & Verreaux, E, 1851)	AF : widespread south of the Sahara
Snow Mountains Quail	Anurophasis monorthonyx (van Oort, 1910)	AU : New Guinea
Jungle Bush Quail	Perdicula asiatica (Latham, 1790)	OR : India, Sri Lanka
	P. a. punjaubi (Whistler, 1939)	nw India
	P. a. asiatica (Latham, 1790)	n and c India
	P. a. vidali (Whistler & Kinnear, 1936)	sw India
	P. a. vellorei (Abdulali & Reuben, 1965)	s India
	P. a. ceylonensis (Whistler, 1936)	Sri Lanka
Rock Bush Quail	Perdicula argoondah (Sykes, 1832)	OR : India
	P. a. meinertzhageni (Whistler, 1937)	w and nw India
	P. a. argoondah (Sykes, 1832)	c and se India
	P. a. salimalii (Whistler, 1943)	s India
Painted Bush Quail	Perdicula erythrorhyncha (Sykes, 1832)	OR : India
	P. e. blewitti (Hume, 1874)	c and se India
	P. e. erythrorhyncha (Sykes, 1832)	sw India
Manipur Bush Quail	Perdicula manipurensis (Hume, 1881)	OR : ne India, Bangladesh
	P. m. inglisi (Ogilvie-Grant, 1909)	n Bengal to n Assam (ne India)
	P. m. manipurensis (Hume, 1881)	Bangladesh to s Assam (ne India)
Himalayan Quail	Ophrysia superciliosa Feared to be extinct. (Gray, 1846)	OR : n India

Excerpt from the IOC World Bird List (2014, version 4.1) editors: Gill, F & D Donsker.

Appendix B: NEW WORLD QUAIL

Order: GALLIFORMES Family: Odontophoridae		
IOC English Name	Scientific Name	Range
Bearded Wood Partridge (Bearded Tree Quail)	*Dendrortyx barbatus* (Gould, 1846)	MA : c Mexico
Long-tailed Wood Partridge (Long-tailed Tree Quail)	*Dendrortyx macroura* (Jardine & Selby, 1828)	MA : c Mexico
	D. m. macroura (Jardine & Selby, 1828)	ec Mexico
	D. m. griseipectus (Nelson, 1897)	c Mexico
	D. m. diversus (Friedmann, 1943)	nw Jalisco (wc Mexico)
	D. m. striatus (Nelson, 1897)	wc Mexico
	D. m. inesperatus (Phillips, AR, 1966)	sc Mexico
	D. m. oaxacae (Nelson, 1897)	w Oaxaca (sc Mexico)
Buffy-crowned Wood Partridge (Buffy-crowned Tree Quail)	*Dendrortyx leucophrys* (Gould, 1844)	MA : s Mexico to Costa Rica
	D. l. leucophrys (Gould, 1844)	s Mexico to Nicaragua
	D. l. hypospodius (Salvin, 1896)	n Costa Rica
Mountain Quail	*Oreortyx pictus* (Douglas, 1829)	NA, MA : sw Canada to Baja California
	O. p. pictus (Douglas, 1829)	sw Washington to nw California (nw USA)
	O. p. plumifer (Gould, 1837)	Oregan, ne California, w Nevada (nw USA)
	O. p. russelli (Miller, 1946)	Little San Bernadino Mts, California. (sw USA)
	O. p. eremophilus (van Rossem, 1937)	s California (sw USA)
	O. p. confinis (Anthony, 1889)	n Baja California (Mexico)
Scaled Quail	*Callipepla squamata*, (Vigors, 1830)	NA, MA : sw US, nc Mexico
	C. s. pallida (Brewster, 1881)	sw USA, nw Mexico
	C. s. hargravei (Rea, 1973)	sc USA
	C. s. castanogastris (Brewster, 1883)	s Texas (USA), ne Mexico
	C. s. squamata (Vigors, 1830)	n and nc Mexico
Elegant Quail	*Callipepla douglasii* (Vigors, 1829)	MA : w Mexico
	C. d. douglasii (Vigors, 1829)	Sinaloa, nw Durango (nw Mexico)
	C. d. bensoni (Ridgway, 1887)	Sonora, c Chihuahua (nw Mexico)
	C. d. teres (Friedmann, 1943)	Nayartie, Jalisco (wc Mexico)

California Quail	Callipepla californica (Shaw, 1798)	NA, MA : sw Canada to Baja California
	C. c. brunnescens (Ridgway, 1884)	sw Oregon to c California (wc USA)
	C. c. canfieldae (van Rossem, 1939)	ec California (wc USA)
	C. c. californica (Shaw, 1798)	e Oregon (USA) to nw Mexico
	C. c. catalinensis (Grinnell, 1906)	Santa Catalina I. (USA)
	C. c. achrustera (Peters, 1923)	s Baja California (nw Mexico)
Gambel's Quail	Callipepla gambelii (Gambel, 1843)	NA, MA : sw US, nw Mexico
	C. g. gambelii (Gambel, 1843)	sw United States, nw Mexico
	C. g. fulvipectus (Nelson, 1899)	Sonora (w Mexico)
Banded Quail	Philortyx fasciatus (Gould, 1844)	MA : c Mexico
Northern Bobwhite	Colinus virginianus (Linnaeus, 1758)	NA, MA : ec, se USA, e Mexico
	C. v. virginianus (Linnaeus, 1758)	s and e USA
	C. v. floridanus (Coues, 1872)	Florida (se USA), Bahama Is.
	C. v. insulanus (Howe, 1904)	Key West, Florida (USA)
	C. v. cubanensis (Gray, 1846)	Cuba
	C. v. taylori (Lincoln, 1915)	c USA
	C. v. ridgwayi (Brewster, 1885)	Sonora (n Mexico)
	C. v. texanus (Lawrence, 1853)	sw Texas (sc USA) to Coahuila, Nuevo-León and Tamaulipas (n Mexico)
	C. v. maculatus (Nelson, 1899)	ec and c Mexico
	C. v. aridus (Aldrich, 1942)	ne Mexico
	C. v. graysoni (Lawrence, 1867)	wc Mexico
	C. v. nigripectus (Nelson, 1897)	e Mexico
	C. v. pectoralis (Gould, 1843)	c Veracruz (se Mexico)
	C. v. godmani (Nelson, 1897)	e Veracruz (se Mexico)
	C. v. minor (Nelson, 1901)	Tabasco, ne Chiapas (se Mexico)
	C. v. insignis (Nelson, 1897)	se Chiapas (s Mexico), nw Guatemala
	C. v. salvini (Nelson, 1897)	s Chiapas (s Mexico)
	C. v. coyoleos (Statius Müller, 1776)	e Oaxaca, n Chiapas (s Mexico)
	C. v. thayeri (Bangs & Peters, 1928)	ne Oaxaca (s Mexico)
	C. v. harrisoni (Orr & Webster, 1968)	sw Oaxaca (s Mexico)
	C. v. atriceps (Ogilvie-Grant, 1893)	w Oaxaca (s Mexico)
Yucatan Bobwhite	Colinus nigrogularis (Gould, 1843)	MA : se Mexico to Honduras
	C. n. caboti (van Tyne & Trautman, 1941)	Campeche (se Mexico)
	C. n. persiccus (van Tyne & Trautman, 1941)	n Yucatan (se Mexico)
	C. n. nigrogularis (Gould, 1843)	Belize, n Guatemala
	C. n. segoviensis (Ridgway, 1888)	e Honduras, ne Guatemala

Spot-bellied Bobwhite	Colinus leucopogon (Lesson, 1842)	MA : Guatemala to Costa Rica
	C. l. incanus (Friedmann, 1944)	s Guatemala
	C. l. hypoleucus (Gould, 1860)	w El Salvador, w Guatemala
	C. l. leucopogon (Lesson, 1842)	se El Salvador, w Honduras
	C. l. leylandi (Moore, TJ, 1859)	nw Honduras
	C. l. sclateri (Bonaparte, 1856)	sw and c Honduras, nw Nicaragua
	C. l. dickeyi (Conover, 1932)	nw and c Costa Rica
Crested Bobwhite	Colinus cristatus (Linnaeus, 1766)	LA : Costa Rica through n SA
	C. c. mariae (Wetmore, 1962)	sw Costa Rica, Chiriqui (nw Panama)
	C. c. panamensis (Dickey & van Rossem, 1930)	sw Panama
	C. c. decoratus (Todd, 1917)	n Colombia
	C. c. littoralis (Todd, 1917)	Santa Marta foothills (ne Colombia)
	C. c. cristatus (Linnaeus, 1766)	ne Colombia, nw Venezuela, Aruba, Curacao
	C. c. horvathi (Madarász, 1904)	Merida Mts. (nw Venezuela)
	C. c. barnesi Gilliard, 1940)	wc Venezuela
	C. c. sonnini (Temminck, 1815)	nc Venezuela, the Guianas, n Brazil
	C. c. mocquerysi (Hartert, 1894)	ne Venezuela
	C. c. leucotis (Gould, 1844)	Magdalena Valley (nc Colombia)
	C. c. badius (Conover, 1938)	wc Colombia
	C. c. bogotensis (Dugand, 1943)	nc Colombia
	C. c. parvicristatus (Gould, 1843)	ec Colombia, sc Venezuela
Marbled Wood Quail	Odontophorus gujanensis (Gmelin, 1789)	LA : Costa Rica to e Bolivia
	O. g. castigatus (Bangs, 1901)	sw Costa Rica, nw Panama
	O. g. marmoratus (Gould, 1843)	e Panama, n Colombia, nw Venezuela
	O. g. medius (Chapman, 1929)	s Venezuela, nw Brazil
	O. g. gujanensis (Gmelin, 1789)	se Venezuela, the Guianas, Brazil, n Paraguay
	O. g. buckleyi (Chubb, 1919)	s and e Colombia, e Ecuador, n Peru
	O. g. rufogularis (Blake, 1959)	ne Peru
	O. g. pachyrhynchus (Tschudi, 1844)	ec Peru, w Bolivia
	O. g. simonsi (Chubb, 1919)	n and e Bolivia
Spot-winged Wood Quail	Odontophorus capueira (von Spix, 1825)	SA : e Brazil to ne Argentina and e Paraguay
Black-eared Wood Quail	Odontophorus melanotis (Salvin, 1865)	MA : Honduras to Panama
	O. m. verecundus (Peters, 1929)	n Honduras
	O. m. melanotis (Salvin, 1865)	se Honduras to Panama

Rufous-fronted Wood Quail	**Odontophorus erythrops** (Gould, 1859) O. e. **parambae** (Rothschild, 1897) O. e. **erythrops** (Gould, 1859)	**SA : w Colombia, w Ecuador** w Colombia, w Ecuador sw Ecuador
Black-fronted Wood Quail	**Odontophorus atrifrons** (Allen, 1900) O. a. **atrifrons** (Allen, 1900) O. a. **variegatus** (Todd, 1919) O. a. **navai** (Aveledo & Pons, 1952)	**SA : ne Colombia, nw Venezuela** Santa Marta Mts (ne Colombia) ne Colombia Sierra de Perijá (ne Colombia, nw Venezuela)
Chestnut Wood Quail	**Odontophorus hyperythrus** (Gould, 1858)	**SA : Colombia**
Dark-backed Wood Quail	**Odontophorus melanonotus**, (Gould, 1861)	**SA : sw Colombia, nw Ecuador**
Rufous-breasted Wood Quail	**Odontophorus speciosus** (Tschudi, 1843) O. s. **soderstromii** (Lönnberg & Rendahl, 1922) O. s. **speciosus** (Tschudi, 1843) O. s. **loricatus** (Todd, 1932)	**SA : w Amazonia** e and s Ecuador ec Peru se Peru, e Bolivia
Tacarcuna Wood Quail	**Odontophorus dialeucos** (Wetmore, 1963)	**LA : e Panama, nw Colombia**
Gorgeted Wood Quail	**Odontophorus strophium** (Gould, 1844)	**SA : Colombia**
Venezuelan Wood Quail	**Odontophorus columbianus** (Gould, 1850)	**SA : nc Venezuela**
Black-breasted Wood Quail	**Odontophorus leucolaemus** (Salvin, 1867)	**MA : Costa Rica, Panama**
Stripe-faced Wood Quail	**Odontophorus balliviani** (Gould, 1846)	**SA : se Peru, nw Bolivia**
Starred Wood Quail	**Odontophorus stellatus** (Gould, 1843)	**SA : w Amazonia**
Spotted Wood Quail	**Odontophorus guttatus** (Gould, 1838)	**MA : s Mexico to Panama**

Singing Quail	**Dactylortyx thoracicus** (Gambel, 1848)	**MA : c Mexico to Honduras**
	D. t. pettingilli (Warner & Harrell, 1957)	se San Luis Potosí and sw Tamaulipas (ec Mexico)
	D. t. thoracicus (Gambel, 1848)	ne Puebla and c Veracruz, ec Mexico (ec Mexico)
	D. t. sharpei (Nelson, 1903)	Yucatan Pen. (se Mexico) to n Guatemala
	D. t. paynteri (Warner & Harrell, 1955)	s Quintana Roo (se Mexico)
	D. t. devius Nelson, 1898)	Jalisco (w Mexico)
	D. t. melodus (Warner & Harrell, 1957)	Guerrero (sw Mexico)
	D. t. chiapensis (Nelson, 1898)	c Chiapas (s Mexico)
	D. t. dolichonyx (Warner & Harrell, 1957)	s Chiapas (s Mexico)
	D. t. salvadoranus (Dickey & van Rossem, 1928)	El Salvador
	D. t. fuscus (Conover, 1937)	c Honduras
	D. t. conoveri (Warner & Harrell, 1957)	e Honduras
Montezuma Quail	**Cyrtonyx montezumae** (Vigors, 1830)	**NA, MA : sw US, Mexico**
	C. m. mearnsi (Nelson, 1900)	sw USA, n and Coahuila (n Mexico)
	C. m. montezumae (Vigors, 1830)	c Mexico
	C. m. merriami (Nelson, 1897)	Mt. Orizaba (ec Mexico)
	C. m. rowleyi (Phillips, 1966)	Oaxaca (s Mexico)
	C. m. sallei (Verreaux, 1859)	Michoacán, Guerrero (sw Mexico)
Ocellated Quail	**Cyrtonyx ocellatus** (Gould, 1837)	**MA : sw Mexico to Nicaragua**
Tawny-faced Quail	**Rhynchortyx cinctus** (Salvin, 1876)	**LA : Honduras to Ecuador**
	R. c. pudibundus (Peters, 1929)	ne Honduras, e and nc Nicaragua
	R. c. cinctus (Salvin, 1876)	s Nicaragua to Panama
	R. c. australis (Chapman, 1915)	w Colombia, nw Ecuador

Excerpt from the IOC World Bird List (2014, version 4.1) editors: Gill, F & D Donsker.

Appendix C: BUTTONQUAIL

Order: CHARADRIIFORMES			
Family: Turnicidae (Buttonquail)			
IOC English Name	Scientific Name	Range
Common Buttonquail	**Turnix sylvaticus** (Desfontaines, 1789)	AF, OR : widespread
	T. s. sylvaticus (Desfontaines, 1789)	s Iberian Pen., nw Africa
	T. s. lepurana (Smith, 1836)	Africa south of the Sahara
	T. s. dussumier (Temminck, 1828)	e Iran to Burma
	T. s. davidi (Delacour & Jabouille, 1930)	c Thailand to s China, n Indochina and Taiwan
	T. s. bartelsorum (Neumann, 1929)	Java, Bali
	T. s. whiteheadi (Ogilvie-Grant, 1897)	Luzon (n Philippines)
	T. s. celestinoi (McGregor, 1907)	Bohol and Mindanao (s Philippines)
	T. s. nigrorum (duPont, 1976)	Negros I. (c Philippines)
	T. s. suluensis (Mearns, 1905)	Sulu Arch. (sw Philippines)
Red-backed Buttonquail	**Turnix maculosus** (Temminck, 1815)	AU : Sulawesi and New Guinea to n Australia
	T. m. kinneari (Neumann, 1939)	Peleng I. (e of Sulawesi)
	T. m. beccarii (Salvadori, 1875)	Sulawesi, Muna and Tomia Is. (Tukangbesi Is.)
	T. m. obiensis (Sutter, 1955)	Kai Is., Babar I.
	T. m. sumbanus (Sutter, 1955)	Sumba I. (Lesser Sundas)
	T. m. floresianus (Sutter, 1955)	Sumbawa, Komodo, Padar, Flores and Alor (Lesser Sundas)
	T. m. maculosus (Temminck, 1815)	Roti, Semau, Timor, Wetar, Moa and Kisar (Lesser Sundas)
	T. m. savuensis (Sutter, 1955)	Sawu I. (Lesser Sundas)
	T. m. saturatus (Forbes, 1882)	New Britain, Duke of York I. (Bismarck Arch.)
	T. m. furvus (Parkes, 1949)	Huon Pen. (ne New Guinea)
	T. m. giluwensis (Sims, 1954)	c New Guinea
	T. m. horsbrughi (Ingram, 1909)	s New Guinea
	T. m. mayri (Sutter, 1955)	Louisiade Arch. (New Guinea)
	T. m. salomonis (Mayr, 1938)	Guadalcanal (Solomon Is.)
	T. m. melanotus (Gould, 1837)	n and e Australia
Hottentot Buttonquail	**Turnix hottentottus** (Temminck, 1815)	AF : South Africa
Black-rumped Buttonquail	**Turnix nanus** (Sundevall, 1850)	AF : widespread

Yellow-legged Buttonquail	**Turnix tanki** (Blyth, 1843)	OR : widespread
	T. t. tanki (Blyth, 1843)	c Pakistan, India, Nicobar and Andaman Is.
	T. t. blanfordii (Blyth, 1863)	se Siberia, Korea and ne China to Burma, Thailand and Indochina
Spotted Buttonquail	**Turnix ocellatus** (Scopoli, 1786)	OR : Philippines
	T. o. benguetensis (Parkes, 1968)	n Luzon
	T. o. ocellatus (Scopoli, 1786)	s and c Luzon
Barred Buttonquail	**Turnix suscitator** (Gmelin, JF, 1789)	OR : widespread
	T. s. plumbipes (Hodgson, 1837)	Nepal to ne India and n Burma
	T. s. bengalensis (Blyth, 1852)	c and s of West Bengal (ne India)
	T. s. taigoor (Sykes, 1832)	India (except above)
	T. s. leggei (Baker, ECS, 1920)	Sri Lanka
	T. s. okinavensis (Phillips, AR, 1947)	s Kyushu I. to Ryukyu Is. (Japan)
	T. s. rostratus (Swinhoe, 1865)	Taiwan
	T. s. blakistoni (Swinhoe, 1871)	e Burma to s China and n Indochina
	T. s. pallescens (Robinson & Baker, ECS, 1928)	sc Burma
	T. s. thai (Deignan, 1946)	nw and c Thailand
	T. s. interrumpens (Robinson & Baker, ECS, 1928)	s Burma and Thailand
	T. s. atrogularis, (Eyton, 1839)	Malay Pen.
	T. s. suscitator, (Gmelin, JF, 1789)	Sumatra, Java, Bawean, Belitung, Bangka and Bali
	T. s. powelli (Guillemard, 1885)	Lombok to Alor (Lesser Sundas)
	T. s. rufilatus (Wallace, 1865)	Sulawesi
	T. s. haynaldi (Blasius, W. 1888)	Palawan and nearby islands (w Philippines)
	T. s. fasciatus (Temminck, 1815)	Luzon, Mindoro, Masbate, Sibuyan (n Philippines)
	T. s. nigrescens (Tweeddale, 1878)	Cebu, Guimaras, Negros, Panay (Philippines)
Madagascan Buttonquail	**Turnix nigricollis** (Gmelin, JF, 1789)	AF : Madagascar
Black-breasted Buttonquail	**Turnix melanogaster** (Gould, 1837)	AU : e Australia
Chestnut-backed Buttonquail	**Turnix castanotus** (Gould, 1840)	AU : n Australia
Buff-breasted Buttonquail	**Turnix olivii** (Robinson, 1900)	AU : n Australia

Painted Buttonquail	**Turnix varius** (Latham, 1802)	AU : sw, se, e Australia
	T. v. **novaecaledoniae** Possibly extinct (Ogilvie-Grant, 1889)	New Caledonia
	T. v. **scintillans** (Gould, 1845)	Houtman Abrolhos Is. (sw of Australia)
	T. v. **varius** (Latham, 1802)	sw, se and e Australia
Worcester's Buttonquail	**Turnix worcesteri** (McGregor, 1904)	OR : Philippines
Sumba Buttonquail	**Turnix everetti** (Hartert, 1898)	AU : Lesser Sundas
Red-chested Buttonquail	**Turnix pyrrhothorax** (Gould, 1841)	AU : n, e, se Australia
Little Buttonquail	**Turnix velox** (Gould, 1841)	AU : Australia

Excerpt from the IOC World Bird List (2014, version 4.1) editors: Gill, F & D Donsker.

Further Information

Website

www.UrbanQuailKeeping.com

On my website you will find lots of useful supplementary material to this book. The material is constantly changing, so be sure to check back from time to time.

Index

A

ACV .... *See* Apple Cider Vinegar
Aggressive male birds ... 48, 49, 96, 104
Air sac 83, 85
Antibiotics 59, 61, 65
Apple Cider Vinegar 64, 67
    Algae prevention 69
    Anti-bacterial 68
    Dosage for quail 70
    Fly & parasite deterrent 69
    How to make 70
    Improving shells 69
    Mother-of-vinegar 71
    Mould & mildew 69
    Odour prevention 68
    Precautions 70
    Worms 68
Avacado (toxic) 43

B

Beaks 52
    Trimming 52, 53
Beans, uncooked (toxic) 43
Bedding 34, 46
    Brooder 92, 93
Behaviour
    Aggression 48
    Bullying 47
    Green vegetables 49
    Hiding places 49
    Pecking *47–50*, 63
    Quiet, withdrawn 51
    Reduce bullying 53
    Sand bath 49
Boredom 48
    Avoiding 48
Breeding *77–102*

Candling 84
Egg storage 80
Fertilized eggs 78, 80
Genetic dominance 101
Growth 93
Handling eggs 80
Incubator 78
Incubator location 82
Lethal combinations 101
Nursery pen 95
Sexual maturity 99
Washing eggs 80
Brooder 78, 86, 88
    Bedding 92, 93
    Cleanliness 94
    D.I.Y. version 88
    Heat source 92
    Temperature 90, 92
    Ventilation 91
Bullying .... *See* Behaviour
Bumblefoot 61

C

Calcium .... See Nutrients
Cannibalism 47
Chick crumb .... *See* Food
Chocolate (toxic) 43
Classification of quail 23
Cleaning out 31, 45
    Toeballing 60
Cloaca 58
Coccidiosis 64, 68, 94
    Enteritis compared 65
Cooking with quail eggs *113–15*
    Boiled egg times 114
    How to crack open 113
    Shell removal 114
Culling *107–10*

Best method 108
Principles of a painless cull .. 108
Tips 110
Cuttlefish bone 36, 49

D

Damp
    Disease control 64
Diarrhoea 51, 64, 65
Diatom Powder 63
Disclaimer iv
Draughts 48, 49
Drinkers .... *See* Water dishes
Drinking water .... *See* Water

E

Egg production *9–18*
Egg recipes *116–27*
Egg tooth 85
Eggs
    Candling 84
    Fertility 80
    How many laid 10
    Odd sized 14
    Quail compared to chicken's .. 11
    Selling, UK regulations 17
    Soft shells 39, 69
    Withdrawal period 65, 66
    Yolk colour 75
Eggshells 75
    Colouring 14
Electric hen 92
Enteritis .... *See* Health

F

F10 barrier ointment 66
Feathers 51
    Apple Cider Vinegar 68
    Colour dominance 101
    Damaged 63
    Male & female differences 96

Mutations from breeding 102
Poor condition 39, 63
Foam balls 99
Food
    Amount per bird 42
    Apples 43
    Chick crumb 39, 92
    Chickweed 42
    Dandelion 42
    Green vegetables 49
    Grit 34, 44
    Pot Marigolds 75
Food dishes 34, 49
    Avoiding contamination 64
    Covered dishes 35
Free range quail 26

G

Genetic dominance .... *See* Feathers, *See* Breeding
Gentian violet spray 66
Grit .... *See* Food

H

Hatching 85
Health *51–66*
    Autopsy 64
    Bumblefoot 61
    Cleanliness 94
    Correct nutrients 41
    Cuts and abrasions 65
    Diarrhoea .... *See* Diarrhoea
    Ulcerative Enteritis 65
    Young chicks 64
Henbane (toxic) 43
Hiding places 36, 49
Holding quail 20
Home craft *67–76*
Hospital cage 52
Housing quail *25–38*, 27
    Aspect / location 30

147

Double layered hutches 31
Flooring 31
Height requirements 31
Hutch designs 30
Keeping dry (avoid damp) 28
Low roof - high roof............... 31
Space per bird......................... 27
Temperature & heating 28
Wind & draughts 28
Humane Slaughter Association 110
Humidity...................................... 82
Hatching 86
Hutches
Hospital cage......................... 52
On legs 31
Runs....................................... 26
Windows 32, 45

I

Incubator............................... 78, 81
Air sac 83, 85
Check list............................... 87
Hatching 85
Humidity 82, 84, 86
Location 82
Pipping 84
Temperature 82, 84
Turning eggs 81

L

Lice 62
Lighting 26
Behaviour 104
Growth rate 106

M

Marigolds *Tagetes* (toxic) 75
Mealworms.................................. 73
Mites (mealworm parasite) 74
Meat
Improving flavour 69

Selling, UK regulations 106
Meat recipes *128–34*
Methionine .... See Nutrients
Mites..48
Diatom Powder 62
Red Mites............................... 61
Scaly Leg 62
Treatment 45
Mother-of-vinegar 67

N

Nail clippers 54, 56
Nature of quail...................... *19–24*
Neighbours and quail 23
Noise.. 23
Nursery pen 95
Nutrients
Adult quail 40
Calcium............................ 41, 76
Methionine 41
Phosphorus............................ 41
Protein.................................... 41
Quail chicks 94

O

Organic 103
Organic food................................ 40
Overcrowding.............................. 48

P

Pecking........................ *47–50*, 63, 65
Aggression 49
Gentian violet spray................ 66
Sick bird treatment................. 52
Pests
Bacteria ....See Bumblefoot
Coccidiosis......... *See* Coccidiosis
Lice .... *See* Lice
Rats 31
Red Mite .... *See* Mites
Scaly Leg (mite) 62

148

Worms (parasite) ...... *See* Worms
Phosphorus .... See Nutrients
Pipping 82, 84, 86
Potato - green, peelings (toxic) ... 43
Preparing for the table *111–12*
Prolapsed vent 58
   Treatment 58
Protein .... See Nutrients

Q

Quail care tasks *44–46*
Quail Disease 65
Quail family tree 23
Quail food *39–43*
Quail for meat *103–6*
   Constant supply calculation . 105
   Selecting for breeding 105
   Selecting to eat 104
   Selling, UK regulations 106
Quiet withdrawn birds 51

R

Rain 48
Recipes
   Almond & polenta herb crusted quail 132
   Christmas ripple ice cream ... 126
   Courgette frittata 120
   Eggs *116–27*
   Eggs in aspic 119
   Eggs with dukkah 118
   Egyptian roast quail 130
   Honey & bourbon marinated quail 131
   Hot chilli pickled quail eggs . 124
   Ice cream 125
   Mini scotch eggs 122
   Muffin in a minute 127
   Pickled quail eggs 123
   Smoked cheese & apple stuffed quail 133

Tandoori quail 134
Thai spiced caramelized eggs 117
Turkish grilled quail 129
Red Mites .... *See* Mites
Response to clothing 23

S

Salt (toxic) 43
Sand bath 36, 45, 46
   Avoiding pecking 49
Scaly Leg 62
   Treatment 62
Sexing quail 96
Sick quail 51
Song of Quail 23
Space requirement 25, 48
Staphylococcus Bacteria .... *See* Bumblefoot
Stressed birds .... *See* Behaviour

T

Taming quail 22
Taste of quail eggs 13
Temperature 44, 48, 49
   Brooder 92
   Chicks 95
   Frost 48
Territorial disputes 48
Toeballing 56, 60, 93
Toenail 52
   Trimming 56
Treats 44
Types of quail 11, 104
   Buttonquail 142
   New World quail 137
   Old World quail 135

U

UK regulations
   Selling eggs 17
   Selling meat 106

V

Ventilation 28, 91

W

Water dishes 36, 49
    Avoiding contamination......... 64
    Brooder................................... 92
    Covered 36

Water, drinking 44
Website...................................... 145
Weight loss 63, 104
Wind 48
Wire mesh 32, 56
Worms (parasitic)............ 45, 63, 68

X

Xanthophyll................................. 75

About The Author

Karen J Puddephatt began quail-keeping many years ago as a means of producing eggs and meat from home. The lack of up-to-date and reliable information for urban quail keepers inspired her to write down the knowledge she had gained from her practical hands-on experience. The resulting book ***Urban Quail-Keeping*** offers straightforward advice which she hopes will inspire you to keep your own flock and reap the benefits from this absorbing hobby.

Printed in Great Britain
by Amazon.co.uk, Ltd.,
Marston Gate.